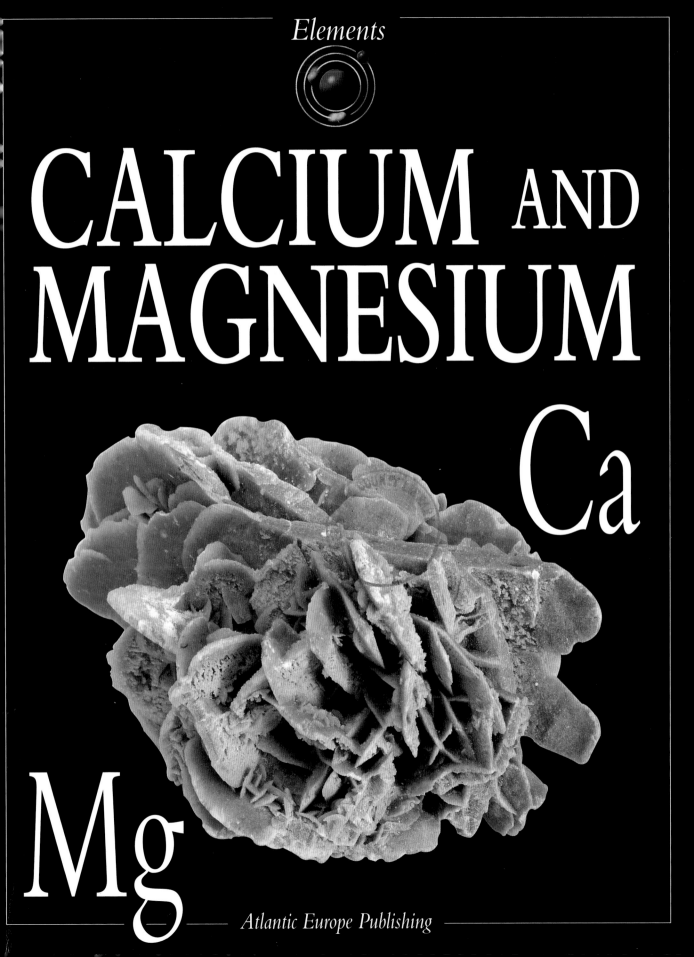

CALCIUM AND MAGNESIUM

Ca

Mg

Atlantic Europe Publishing

How to use this book

This book has been carefully developed to help you understand the chemistry of the elements. In it you will find a systematic and comprehensive coverage of the basic qualities of each element. Each two-page entry contains information at various levels of technical content and language, along with definitions of useful technical terms, as shown in the thumbnail diagram to the right. There is a comprehensive glossary of technical terms at the back of the book, along with an extensive index, key facts, an explanation of the Periodic Table, and a description of how to interpret chemical equations.

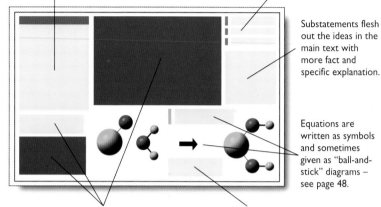

The main text follows the sequence of information in the book and summarises the concepts presented on the two pages.

Technical definitions.

Substatements flesh out the ideas in the main text with more fact and specific explanation.

Equations are written as symbols and sometimes given as "ball-and-stick" diagrams – see page 48.

Photographs and diagrams have been carefully selected and annotated for clarity.

Also… explains advanced concepts.

An Atlantic Europe Publishing Book

Author
Brian Knapp, BSc, PhD
Project consultant
Keith B. Walshaw, MA, BSc, DPhil
(Head of Chemistry, Leighton Park School)
Industrial consultant
Jack Brettle, BSc, PhD (Chief Research Scientist, Pilkington plc)
Art Director
Duncan McCrae, BSc
Editor
Elizabeth Walker, BA
Special photography
Ian Gledhill
Illustrations
David Woodroffe
Electronic page make-up
Julie James Graphic Design
Designed and produced by
EARTHSCAPE EDITIONS
Print consultants
Landmark Production Consultants Ltd
Reproduced by
Leo Reprographics
Printed and bound by
Paramount Printing Company Ltd

Suggested cataloguing location
Knapp, Brian
   Calcium and magnesium
   ISBN 1 869860 14 4
   – Elements series
540

Acknowledgements
The publishers would like to thank the following for their kind help and advice: *Jonathan Frankel* of J.M. Frankel and Associates, *Ian* and *Catherine Gledhill* of Shutters, *John Chrobnick, Steve Rockell* of Goodrock Properties Services Ltd, *Pippa Trounce* and *Mike L. Willoughby.*

Picture credits
All photographs by **Earthscape Editions**.
Front cover: A piece of fossil-rich oolitic limestone. Limestone is mostly white calcium carbonate but in this case is reddish due to iron staining and some contamination by mud.
Title page: Desert roses are crystals of a compound of calcium called calcium sulphate (gypsum).

This product is manufactured from sustainable managed forests. For every tree cut down at least one more is planted.

WEST GRID STAMP
NN
NT
NC
RS
NH
NL
NV
NM
ND
NE
NP

First published in 1996 by
Atlantic Europe Publishing Company Limited, Greys Court Farm,
Greys Court, Henley-on-Thames, Oxon, RG9 4PG, UK.

Copyright © 1996
Atlantic Europe Publishing Company Limited

The demonstrations described or illustrated in this book are not for replication. The Publisher cannot accept any responsibility for any accidents or injuries that may result from conducting the experiments described or illustrated in this book.

Contents

Introduction 4

Calcium 6

Crystals containing calcium 8

Limestone 10

Dissolving limestone 12

Caves and caverns 14

Calcium carbonate 16

Building stone 18

Calcium oxide 20

Using calcium oxide 22

Calcium hydroxide 24

Limewater 26

Calcium in the soil 28

Calcium sulphate 30

Bones 32

Magnesium 34

Protecting with magnesium 36

Hard water 38

Softening water 40

Antacids 42

Key facts about calcium and magnesium 44

The Periodic Table 46

Understanding equations 48

Glossary of technical terms 50

Index 56

Introduction

An element is a substance that cannot be broken down into a simpler substance by any known means. Each of the 92 naturally occurring elements is therefore a fundamental substance from which everything in the Universe is made. This book is about calcium and magnesium.

Calcium

Probably everyone has watched a teacher use a stick of "chalk" to mark a blackboard. Blackboard chalk is a calcium-rich material. It is a compound containing the element calcium (symbol Ca) bonded together with one other element or more. Calcium on its own is actually a soft, silvery-coloured metal. However, calcium only occurs in nature as part of compounds.

The most common calcium compound is a material called calcium carbonate. Most people know calcium carbonate as rock, such as chalk, limestone and marble. They may have admired coral reefs without realising that both the coral rock and the tiny green plants that live with the coral, known as algae, have calcium in them. Calcium even contributes to the pearls that are used to make rings and necklaces.

When gardeners use crushed limestone, crushed bone meal, or pulverised shell as a soil-improver in their gardens, few may realise that the bones of animals as well as the shells of snails, mussels and myriad other creatures all contain calcium.

The bones of people also have calcium compounds in them. In fact, about one-fiftieth of the human body is made from calcium. Some uses you may be able to guess, such as the formation of bones and teeth, but others are quite invisible to us, such as making our muscles move!

But the element calcium is not just in natural things. For example, it is also in the walls of the house around you. Its compounds make up the plaster used in walls, the cement that holds bricks firmly together and concrete that makes the base for the house.

Magnesium

In many ways magnesium (symbol Mg) is very similar to calcium. Magnesium is one of the most important metals in the body, making up no less than one-fortieth of each and every one of us. It is also a very important element in the make-up of all green plants.

Like calcium, magnesium is never naturally found alone but always as a compound combined with other elements. For example, it makes up the hard limestone-looking rocks of magnesium carbonate called dolomite. Magnesium carbonate is used as a filler in paper and as an antacid. And because magnesium burns with an intense white light, it is also useful as a signal flare.

This book will introduce you to some of the wide range of properties of calcium, magnesium and some of the compounds they form. It is a starting point for your exciting exploration into the world of chemistry.

▲ Calcium metal burns with a brick-red flame. The flame test is an important way to look for the presence of calcium.

Calcium

Calcium is a soft silvery-coloured metal that reacts with other elements to form compounds. In fact, it reacts so easily that it is never found on its own in nature.

If you see calcium metal in a laboratory, it will often be in the form of small pieces called pellets. This is to make it easy to use in experiments like the one shown on these pages. With pellets it is possible to demonstrate how calcium reacts with one of the most common compounds of all – water.

❶▼ Calcium rapidly reacts with oxygen in the air to form a protective (dull) coating that tends to prevent any further reaction. It only looks silvery when freshly cut.

❷▼ A small pellet of calcium is placed in a beaker of water. At first it sinks, showing that it is heavier than water.

A test tube filled with water is placed over the pellet. Notice that bubbles of hydrogen gas are already rising through the water. This is a result of the reaction of calcium with water.

These calcium pellets look dark because of a surface coating of calcium oxide.

A calcium pellet is placed in a beaker of water and under a test tube.

❸▼ The calcium pellet rises up the test tube, buoyed by bubbles of hydrogen gas.

The solution in the test tube quickly becomes saturated with calcium hydroxide from the reaction. Calcium hydroxide now forms tiny particles (precipitate), making the water appear cloudy.

The hydrogen gas released by the vigorous reaction between the calcium pellet and the water soon fills the test tube.

The cloudiness in the water is caused by particles of calcium hydroxide.

❹▲ On reaching the surface the bubbles burst and the calcium pellet again sinks down into the beaker. Each time the reaction becomes more violent and the bubbles become bigger. Big bubbles like this can form when the oxide coating has been removed, allowing the metal to react more rapidly. The reaction also gives out heat. The precipitate of calcium hydroxide is granular, remaining suspended in the water for some time and making it cloudy.

❺▲ The calcium bobs up and down in the water for a while and then stays on the surface because the calcium pellet gets so small it is more and more easily lifted to the surface by bubbles.

The tube fills with hydrogen and all the water is pushed out of the tube. This stops the reaction.

EQUATION: Calcium in water

Calcium + water ⇨ hydrogen gas + calcium hydroxide

$$Ca(s) \quad + \quad 2H_2O(l) \quad ⇨ \quad H_2(g) \quad + \quad Ca(OH)_2(aq)$$

(calcium hydroxide solution is also known as limewater)

Crystals containing calcium

The most common compound of calcium is limestone, calcium carbonate. This is often a dull grey rock, but occasionally, in small cavities in the rocks, it makes crystals, and the true brilliance of the mineral shows through. The crystalline form of calcium carbonate is called calcite.

Calcite can be found all over the world. It often occurs along with rare metals. In fact, because it is so common, and has such a sparkling white colour, it is very easy to spot. It has led prospectors to find such metals as gold, tin, silver and copper.

Forms of calcite

There is a pure, completely transparent version of calcite. It is known as Iceland Spar because it is sometimes found in cavities in solidified lava, of which Iceland is made. When you look into a piece of Iceland Spar you actually see double because it has the wonderful property of showing two images of anything you see through it.

More often, calcite forms sparkling white crystals, which are sometimes found on the surface near hot springs. Geologists call this material travertine (see page 16). It was the most widely used building stone of ancient Rome.

Other calcium-rich crystals

Calcium compounds can also make crystals known as "Blue John", a corruption of the French words *bleu* and *jaune* for blue and yellow. Geologists call it fluorspar (calcium fluoride) and it makes beautiful green, blue and yellow cubic crystals.

▼ These are desert roses, a compound of calcium called calcium sulphate (also called gypsum). The crystals of this mineral look a bit like rose petals, hence its name. It has a different look from the more block-like rhombic crystals of calcite.

◀ This fossil ammonite has been cut in half so that you can see how its original form has been replaced by calcite. The calcite appears as the grey in-filling to the shell segments. In a few cases you can see how the calcite has formed in small cavities, giving delicate crystals.

lava: the material that flows from a volcano.

mineral: a solid substance made of just one element or chemical compound. Calcite is a mineral because it consists only of calcium carbonate, halite is a mineral because it contains only sodium chloride, quartz is a mineral because it consists of only silicon dioxide.

prospector: a person who is exploring for geologically rich deposits of metals and gemstones.

▶ This rhombohedral-shaped crystal is typical of calcite. The faces are parallelograms.

▶ Crystals of calcite growing in a small cavity in a limestone rock. Notice the limestone fossils. There are more examples of fossils cast in limestone on the next page.

Limestone

Limestone is the common name for rocks that are made up mainly of calcium carbonate. Limestones vary greatly and have many origins.

The origins of limestone

Most limestone rocks share an origin in ancient warm, shallow seas. Some were formed as vast coral reefs. Others were formed from the cemented remains of tiny sea creatures and made into soft chalk. The warm waters also caused calcium to precipitate out from sea waters rather like the scale in a kettle. This scale took the form of millions of tiny balls of limestone. This type of limestone is called oolitic limestone after the Greek word *oon*, meaning egg-shaped.

Water-bearing rocks

Most limestones have cracks and gaps that make them porous. Oil and water can then accumulate in these gaps. This is what makes limestone good oil-bearing and water-bearing rock.

The colour of limestone

Limestone is rarely white because of the impurities in it. It is most commonly light grey, a result of a mixture of calcite and mud. Oolitic limestone is often honey-coloured because it contains some iron. Limestones can contain so much iron that they are worth mining as iron ore.

▲ This is a quarry in Portland. Limestone is one of the world's most famous building stones.

coral reef: a region of the sea-bed where corals grow in massive banks.

ore: a rock containing enough of a useful substance to make mining it worthwhile.

porous: a material containing many small holes or cracks. Quite often the pores are connected, and liquids, such as water or oil, can move through them.

▼ Bands of limestone weather differently when exposed to the air, sometimes creating spectacular landscapes called karst scenery.

▲ A piece of fossil-rich limestone. The body of the limestone is made up of tiny limestone balls (it is oolitic limestone). The reddish colouring is due to iron staining.

◀ This limestone is made up of corals that became engulfed in a grey, lime-containing mud.

Dissolving limestone

Although calcium carbonate (limestone) will not dissolve in pure water, it will react easily with an acid. The reaction happens naturally as acidic rainwater seeps through soils and reaches limestone rocks.

Carbon dioxide, a gas found naturally in the atmosphere, dissolves in raindrops and produces carbonic acid. Its effect on limestone is slow but unceasing. More carbon dioxide is produced in the tiny passageways of soil. This dissolves in the water, seeps down to rocks and causes underground limestone to react and dissolve faster than surface rock.

This slow natural solution of the rock is called weathering. Sometimes the surface soil is stripped off a limestone rock and you can see the way the joints have been widened by chemical weathering. In some cases the joints are widened enough to produce holes big enough to swallow entire rivers.

▼ **Swallow holes**

Swallow holes, sometimes called sink holes, are depressions in the surface of limestone that have been produced by extreme dissolution of the limestone blocks.

As the blocks dissolve away, they are no longer able to support each other, and they collapse.

Swallow holes may be the entrances to entire cave systems, and rivers may disappear into them. This example is in the Tarn Valley in southern France.

EQUATION: Dissolving limestone

Water + carbon dioxide ⇨ carbonic acid

$$H_2O(l) \quad + \quad CO_2(g) \quad ⇨ \quad H_2CO_3(aq)$$

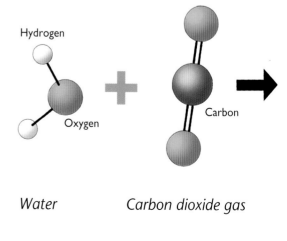

Hydrogen

Oxygen

Carbon

Water

Carbon dioxide gas

dissolve: to break down a substance in a solution without a resultant reaction.

solution: a mixture of a liquid and at least one other substance.

weathering: the slow natural processes that break down rocks and reduce them to small fragments either by mechanical or chemical means.

The faster effects of pollution

Weathering is an example of a general chemical process called corrosion. But in urban areas, and those suffering from acid rain, carbon dioxide is not the only gas in the rainwater. Sulphur dioxide and several oxides of nitrogen are also present. These pollutants are produced by burning fossil fuels. They produce a more concentrated acid, and the weathering effect is much more rapid.

Corrosion produced by acid rain has destroyed many of the limestone sculptures on famous buildings such as ancient cathedrals.

◀ This limestone sculpture on Reims cathedral in France is corroding because of pollution gases in the rainwater.

EQUATION: Dissolving limestone

Carbonic acid + calcium carbonate ⇨ calcium bicarbonate

$$H_2CO_3(aq) \quad + \quad CaCO_3(s) \quad \Rightarrow \quad Ca(HCO_3)_2(aq)$$

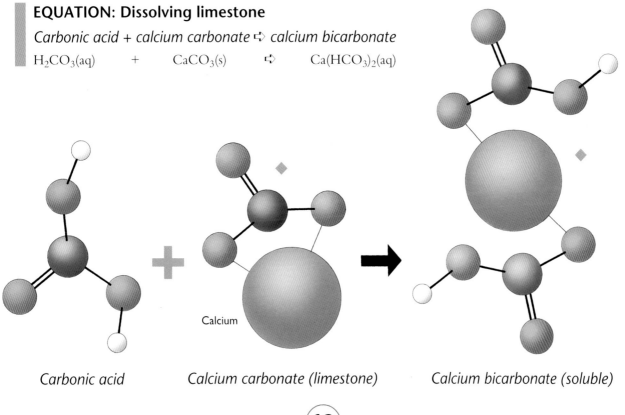

Calcium

Carbonic acid   *Calcium carbonate (limestone)*   *Calcium bicarbonate (soluble)*

13

Caves and caverns

Calcium-containing materials are precipitated in many forms in caves and passages underground. They may form sheets of stone that mask the cave walls, rising columns called stalagmites, or descending columns called stalactites. However, each is made as water laden with calcium carbonate seeps into the caves.

How water becomes rock

The calcium in hard water cannot be seen as it percolates out of tiny cracks in the walls and roofs of the cave. The water looks entirely clear. This is because the calcium is contained as calcium bicarbonate, a transparent solution.

As the water enters the cave, some of the dissolved carbon dioxide escapes, and as a result calcium carbonate is precipitated as tiny crystals of calcite, known as travertine. Over time these build into the fantastic forms inside some caves.

Cave formations

The shape that grows depends on the way the water drips from the roof and walls. If it drains over the walls, then curtain stalactites or flowstone curtains will form. But if it drips vertically then stalactites will form on the roof of the cave and stalagmites will form directly below, at the point where the water drips onto the floor.

Stalactites are not solid, but rather consist of conical tubes. The water flows down the centre of each tube and deposits more crystals on the outside edges of the drip. Stalagmites are thicker and more rounded in shape than stalactites and are solid rather than hollow.

The chemical precipitation takes place very slowly. In fact, it is uncommon for a stalactite to grow at rate greater than two millimetres a year. Stalactites that are many metres long are therefore very old.

◄ Stalagmites, Carlsbad Caverns, New Mexico, USA.

▼ Small wind-distorted stalactites, Caverns of Sonora, Texas, USA.

mineral-laden: a solution close to saturation.

percolate: to move slowly through the pores of a rock.

◄ When water drips out of cracks in a cave, some of the carbon dioxide gas escapes from the water and the calcium bicarbonate changes to calcium carbonate (insoluble limestone). This is precipitated as a tiny addition to the cave. This picture shows a dripping limestone stalactite, Jenolan Caves, Australia.

EQUATION: Precipitating limestone

Calcium bicarbonate ⇨ carbon dioxide + water + calcium carbonate

$$Ca(HCO_3)_2(aq) \quad ⇨ \quad CO_2(g) \quad + \quad H_2O(l) \quad + \quad CaCO_3(s)$$
solution — precipitate

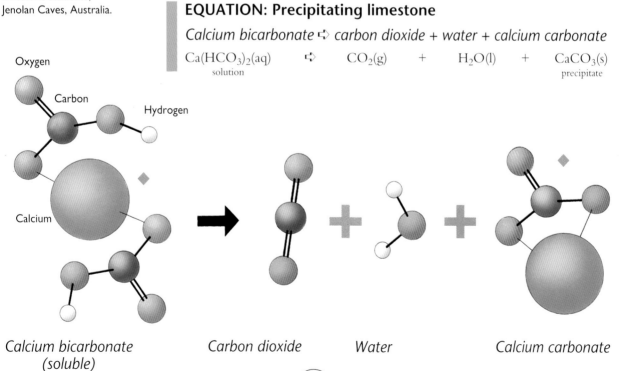

Oxygen

Carbon

Hydrogen

Calcium

Calcium bicarbonate (soluble)

Carbon dioxide

Water

Calcium carbonate

Calcium carbonate

Because calcium bicarbonate is far less stable in hot water than in cold, just like the limescale on a kettle, calcium carbonate (travertine) is often precipitated around many hot springs.

The water in hot springs may have begun as pure rainwater, but by the time it has circulated through underground passages it has dissolved much of the rock through which it is passing and thus has picked up considerable amounts of calcium compounds in solution. As this hot water reaches the ground it cools and the calcium carbonate is precipitated as sheets of white calcite crystals that sparkle in the sunshine.

Waterfalls that sparkle

In the most spectacular locations, such as Pammukale in Turkey and Mammoth Hot Springs in Wyoming, USA (shown here), calcite makes beautiful stepped pools of travertine.

The streaks of other colours in the travertine are formed by other minerals or mud. The pools are enclosed by rimstone dams. These are formed as the water flows over the rim where it is more likely to evaporate. This leads to a build-up of carbonate precipitate, which increases the height of the dam wall.

Calcium sinter

Geysers are dramatic forms of hot springs, sending out gushes of hot, mineral-rich water, sometimes containing calcite, on other occasions silica-rich minerals. As the wind catches the spray of these natural fountains, it may wash over debris and coat it with hardened calcite, forming a sinter. In this way pieces of twig or other small objects can be fossilised.

calcite: the crystalline form of calcium carbonate.

precipitate: tiny solid particles formed as a result of a chemical reaction between two liquids or gases.

stable: able to exist without changing into another substance.

▼ Mammoth Hot Springs, Yellowstone National Park, Wyoming, USA.

Building stone

The most common form of building stone that uses calcium-rich material is made of calcium carbonate, usually in the form of limestone. Many limestones are easy to cut up into blocks and so make good building stones. Calcium carbonate is also found as marble.

Limestone

Limestone is used the world over by architects to make buildings. It is reasonably hard, quite attractive, and yet can be easily sculpted and cut in any direction. Limestone comes from quarries where thick bands of rock are found.

Sometimes limestones are chosen for the fossils they contain. When cut and polished, the fossils make interesting decorations.

Limestone will react with rainwater and weather over a period of time. This can readily be seen in many old buildings.

Marble

Marble is limestone that has been naturally altered by millions of years in the depths of ancient mountains. The impurities show up as the mottling and streaks: red for iron, blue for graphite (a form of carbon) and so on. The streaks were formed when the marble became so hot that the impurities began to melt and run through the limestone.

Marble is much harder than limestone, and more difficult to work with. It is used as the facing for many buildings, and for some of the world's finest sculptures. Buildings containing huge amounts of marble include the White House in Washington DC and the Taj Mahal in India.

▲ The Capitol Building, Washington, DC, USA, uses marble cladding to provide a striking appearance.

▲ The Taj Mahal, India, the world's most famous marble-clad building.

▼ A piece of marble.

facing: sheets of stone cut to form the decorative outer surface of a building. It is a form of cladding. On modern buildings it is also sometimes called siding.

weather: a term used by Earth scientists and derived from "weathering", meaning to react with water and gases of the environment.

▲ Vezelay Abbey, France. Notice how rainwater containing carbon dioxide has begun to dissolve the limestone. The effect is most noticeable on the corners of each block, giving each of the blocks a more rounded appearance.

▶ The Radcliffe Camera, Oxford University, England, a building made from a cream-coloured limestone.

Calcium oxide

Calcium oxide, also known as quicklime, is a major constituent of every bag of cement you buy.

If limestone (calcium carbonate) is heated in a kiln it will decompose to produce carbon dioxide gas and create a white solid of calcium oxide. A laboratory demonstration of the effect is shown here.

❶▼ The apparatus below shows how a small block of calcium carbonate (limestone) can be made into calcium oxide in the laboratory. It consists of an iron cylinder (the kiln) with a tray on which the limestone rests.

❷▶ A cover is put over the kiln and the limestone heated using a Bunsen burner. The limestone decomposes to form a white solid, called calcium oxide or quicklime, and gives off carbon dioxide gas. In the picture you can see the limestone glowing a yellow–red colour.

base: a compound that may be soapy to the touch and that can react with an acid in water to form a salt and water.

decompose: to break down a substance (for example by heat or with the aid of a catalyst) into simpler components. In such a chemical reaction only one substance is involved.

kiln: a structure designed for baking minerals. Lime, brick and pottery kilns are all common.

Rotary kiln (red tube)

◀▲ In a cement works one of the larger pieces of equipment is an iron tube, slightly tilted down at one end and slowly rotating. Inside, limestone is roasted using a gas flame. This equipment, known as a rotating kiln, is the industrial equivalent of the process shown opposite.

EQUATION: Calcium carbonate decomposes

Calcium carbonate ⇨ calcium oxide + carbon dioxide

$$CaCO_3(s) \rightleftharpoons CaO(s) + CO_2(g)$$

❸◀ This picture shows what calcium oxide looks like after it has been cooled and then removed from the kiln.

Also...

Calcium oxide is a strong base and can be used to neutralise acids. Its violent reaction with water makes it unsuitable to be used directly as a soil conditioner (see calcium hydroxide, page 24). However, in dry form it can be used in the manufacture of cement, mortar and concrete.

Calcium carbonate breaks down when heated to form calcium oxide and releases carbon dioxide gas.

The technical word for this kind of breakdown is "thermal decomposition". However, this reaction is reversible. If carbon dioxide gas flows over calcium oxide, it will re-form calcium carbonate (although it will never again look like the original rock).

Using calcium oxide

Lime is a greyish–white powder made by heating limestone. Lime (also called burnt lime and quicklime, see page 20) is one of the most important chemicals known, being sixth after salt, coal, sulphur, air, and water in the amounts used in our world. It is used as a foundation for making many other chemicals.

Mortar

Lime has been used since early times as a simple form of cement known as a mortar. When lime reacts with water, it gives off heat and changes to a new material that is an adhesive (glue). As the water evaporates, the mortar hardens.

Over time carbon dioxide gas, a natural part of the atmosphere, reacts with the mortar, turning it back into the calcium carbonate from which it was originally made. It then falls away as a white powder.

Glass

Lime is used in the making of glass, where it adds hardness and makes the glass insoluble.

Burials and compost heaps

Lime is also a nasty substance to deal with in its pure form. It is caustic, can burn the skin and cause trouble if the powder is breathed in. Lime was traditionally used in burials to hasten the decomposition of bodies.

▲ In the interests of economy these Balinese Geruda images have been cast in concrete, although they used to be carved from basalt.

◄ Whitewash paint
This old house is made of natural boulders. It has been painted with whitewash, a mixture of lime and water.

Because lime picks up carbon dioxide from the air, it eventually turns back into chalk. Thus, the surface of whitewashed walls is often dusty.

Whitewash is cheap, but it has to be re-applied more often than modern paints.

▼ Most apartments are now made almost entirely with concrete. These are in Hong Kong.

Cement: calcium "glue"

One of the main uses of calcium compounds is in the manufacture of cement. Like mortar, cement is an adhesive (glue) used to bond bricks together or to bind stones and gravel to make concrete. Most cement is called Portland cement, a general kind of cement mix named after the English limestone (called Portland Stone) that was used in the cement patented by Joseph Aspdin in 1824.

Making cement

Cement is made from a combination of a limestone or chalk and a clay. The raw materials are ground down and then mixed together. Once mixed they are roasted in a kiln that constantly rotates. The roasting temperature is very high, about the same as that used to melt glass (1350°C). The material is then cooled and crushed to make a fine grey powder.

Using cement

Cement powder is mixed with water to make cement, a grey, pasty substance. The chemical reaction that takes place happens very quickly, and if nothing were done to slow down the process, the cement would go hard in a few minutes and so be very difficult to use. The key to slowing down the reaction is to add gypsum (calcium sulphate).

Cement will last for many decades, but it is not as long-lasting as most other building materials. When exposed to the weather, it will eventually return to the calcium carbonate from which it was made.

Calcium hydroxide

Calcium hydroxide is a white solid. You may have seen cartons containing it in a garden shop, under the name of "slaked lime" or "hydrated lime", where it is sold as a soil conditioner.

Calcium hydroxide can be formed by adding water to calcium oxide. The effect of this is shown on these pages. It is a reaction that gives out considerable amounts of heat.

Another way of obtaining calcium hydroxide is shown on page 7, but calcium is too expensive to obtain as a metal for widespread industrial use.

❶▼ In the demonstration on this page, a block of calcium oxide (quicklime) is placed on a tray. Water is now poured onto one side of the block. This is so that changes that happen to the side in contact with water can then be compared to the original state of the calcium oxide, in its dry form.

The reaction is spectacular. The water soaks into the calcium oxide and disappears. The part of the block of quicklime in contact with the water gets extremely hot, swells and gives off steam. (There is a picture of the original solid on page 21.)

EQUATION: The formation of calcium hydroxide

Calcium oxide + water ⇨ calcium hydroxide

$$CaO(s) \quad + \quad H_2O(l) \quad ⇨ \quad Ca(OH)_2(aq)$$

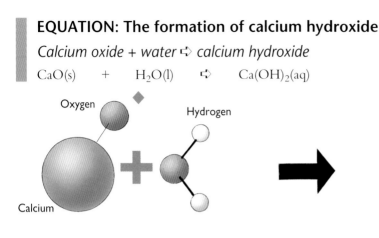

Calcium oxide   Water

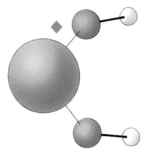

Calcium hydroxide

base: a compound that may be soapy to the touch and that can react with an acid in water to form a salt and water.

ion: an atom, or group of atoms, that has gained or lost one or more electrons and so developed an electrical charge.

❷▼ The block of quicklime expands and begins to crack. Eventually it collapses to a "dry" powder, calcium hydroxide.

Also...

Calcium hydroxide is a strong base and can be used to neutralise acids. It can thus be spread onto an acid soil to neutralise it and provide better conditions for plant growth. Some of the calcium hydroxide breaks apart in water to yield a solution of positively charged calcium ions (Ca^{2+}) and negatively charged hydroxide ions (OH^-). In this state the calcium ions can cause soil particles to stick together (flocculate). Calcium ions can also be taken up in water present in soil and used by plants for growth.

Limewater

In the demonstration on pages 6 and 7, you will have noticed that, as the calcium pellet becomes smaller, the water becomes more cloudy. This is because the reaction with water produces a new solution, known as limewater – calcium hydroxide.

When limewater is saturated, calcium hydroxide is precipitated. A precipitate can be made to appear and disappear at will, as the following demonstration shows.

◄ Solution created by reacting a calcium pellet with water (see pages 6 and 7) is passed down a glass rod through a filter paper in a funnel. A clear solution of calcium hydroxide passes out of the funnel and is collected in the beaker.

❶► As the calcium pellet reacts, it forms calcium hydroxide from the water. Calcium hydroxide is not very soluble in water. If the calcium pellet is large, the solution will contain as much calcium hydroxide as possible: it will be saturated. As more calcium reacts, it will not be possible for more calcium hydroxide to be kept in solution and so the surplus is made into tiny solid particles. This change from solution to solid is called precipitation. The suspension of these tiny particles in the liquid is what causes it to become cloudy. The particles eventually settle out on the bottom of the beaker, leaving the solution clear. They can be filtered from the solution, leaving a clear liquid.

❷► Blowing a stream of bubbles into limewater using a straw produces a startling effect. While breathing through the straw, the liquid appears milky.

Breath contains carbon dioxide gas. This reacts with the calcium hydroxide solution, producing calcium carbonate, which is much less soluble in water. The calcium carbonate is seen as tiny white particles and so the liquid becomes cloudy.

EQUATION: Limewater to calcium carbonate

Calcium hydroxide + carbon dioxide ⇨ calcium carbonate + water

$$Ca(OH)_2(aq) \quad + \quad CO_2(g) \quad ⇨ \quad Ca(CO)_3(s) \quad + \quad H_2O(l)$$

❸▼ If you keep blowing through the straw, the solution quickly goes clear again because the additional amount of carbon dioxide reacts with the calcium carbonate to produce soluble, and colourless, calcium bicarbonate, just like the rainwater and limestone described on page 13.

described on page 13.

precipitate: tiny solid particles formed as a result of a chemical reaction between two liquids or gases.

solution: a mixture of a liquid and at least one other substance (e.g. salt water). Mixtures can be separated out by physical means, for example by evaporation and cooling.

EQUATION: Calcium carbonate to calcium bicarbonate

Calcium carbonate + water + carbon dioxide ⇨ calcium bicarbonate

$$CaCO_3(s) \quad + \quad H_2O(l) \quad + \quad CO_2(g) \quad ⇨ \quad Ca(HCO_3)_2(aq)$$

❹▶ Boiling the colourless solution will turn the solution cloudy again because calcium bicarbonate is not stable in hot water and the bicarbonate changes back to calcium carbonate, which is insoluble.

EQUATION: Change from calcium bicarbonate solution to calcium carbonate

Calcium bicarbonate solution ⇨ calcium carbonate + carbon dioxide + water

$$Ca(HCO_3)_2(aq) \quad \underset{\text{with heat}}{⇨} \quad CaCO_3(s) \quad + \quad CO_2(g) \quad + \quad H_2O(l)$$

Calcium in the soil

Calcium compounds are used in a soil to improve its condition and to balance any acidity. This means keeping the soil materials clumped together into crumb-sized pieces so that the soil drains well and air can get in. For this purpose calcium compounds are used as a form of "glue", holding the fine clay particles clumped into the size of sand grains.

Several compounds of calcium are used for this purpose. The fastest-acting compound is calcium hydroxide (slaked lime, often just called lime by gardeners), which readily breaks down in the soil. A more slow-acting compound is crushed limestone, calcium carbonate.

Lime is not always the best choice for a soil conditioner, however, because it will make a soil alkaline, and plants do not like too alkaline conditions any more than they like acid conditions. This is why it is now more common to apply finely crushed limestone. The limestone will simply remain in the soil until it is dissolved by acid water. In effect the crushed limestone is ready to neutralise the soil when needed but does not cause any damage when not required.

▶ This picture is an illustration of what happens in soil at a microscopic level. The small balls represent calcium ions that are attracted to pieces of clay by electrical charges (see "Also..." opposite). There is always a balance between the calcium ions in the water and those on the soil surfaces. These calcium ions can also be sucked up by plants with soil water.

EQUATION: Neutralisation in soil

Hydroxide ions + hydrogen ions ⇨ water

$$OH^-(aq) \quad + \quad H^+(aq) \quad ⇨ \quad H_2O(l)$$

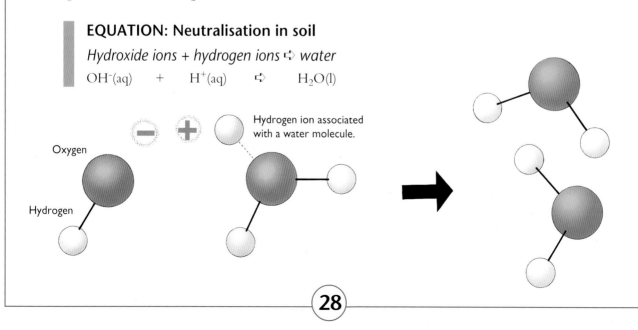

Oxygen

Hydrogen

Hydrogen ion associated with a water molecule.

acidity: a general term for the strength of an acid in a solution.

anion: a negatively charged atom or group of atoms.

cation: a positively charged atom or group of atoms.

ion: an atom, or group of atoms, that has gained or lost one or more electrons and so developed an electrical charge.

◄ Farmers sometimes apply calcium compounds to their fields to help the properties of the soil.

▼ This is a microscope picture of a soil. Notice how there are large gaps (light areas) between the soil particles (dark objects). These pores are where water is stored.

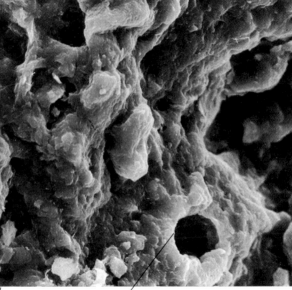

Clay particles clumped, or flocculated, by calcium ions.

Root hair

Calcium ions

Also...

Calcium hydroxide exists as two types of charged particle (known as ions): calcium ions (which are positively charged, or cations) and hydroxide ions (which are negatively charged, or anions). The hydroxide ions combine with any hydrogen ions from acids in the soil to form water. In this way they remove any acidity. The calcium ions attach themselves to the surfaces of fine clay particles in the soil (which have a negative charge). (It is the same sort of effect as rubbing a balloon so that it sticks to your clothes.) Calcium ions have two charges and so can act as links between two clay particles. In this way they bind the soil together, a process called flocculation.

Calcium sulphate

Plaster is made of another important compound of calcium, calcium sulphate.

This material, used to make plaster, occurs naturally as a soft white rock known as gypsum. It can be made into a fine form and used as a crack filler or poured into moulds to make plaster figures. This material is called Plaster of Paris.

Where gypsum is found

Gypsum is found in thick beds, often in the same place as rock salt. This is because gypsum was formed by the evaporation of sea water in ancient seas. It is still being formed in some inland salt lakes, like Australia's Lake Eyre and Utah's Great Salt Lake. Gypsum is cheap to quarry and so is a good material to use in house building, where large amounts of inexpensive materials are needed.

Uses of plaster

Because gypsum is easy to make into a paste and will set hard as it dries, it is used throughout the world as wallboards. The gypsum is put in shallow moulds and then covered with a sheet of strong paper. This stops the gypsum board breaking as it is carried about.

Gypsum wallboards are soft and easy to cut into shapes. They also give the very smooth finish that we expect of the walls in our rooms.

▶ Plaster (gypsum and water made into a paste) can be applied with a trowel to finish off a wall. The main wall surfaces are gypsum too, in the form of prefabricated wallboards.

dehydration: the removal of water from a substance.

hydration: the absorption of water by a substance. Hydrated materials are not "wet" but remain firm, apparently dry, solids. In some cases, hydration makes the substance change colour, in many other cases there is no colour change, simply a change in volume.

◀ This pictures shows a piece of crystalline gypsum, calcium sulphate. Gypsum is soft enough to be scratched by a fingernail.

Also...

Plaster of Paris is a special form of the salt calcium sulphate. It is made by heating crushed gypsum to drive off some of the water. It is then crushed to a fine powder. When Plaster of Paris is mixed with water, the calcium sulphate takes up water again (it hydrates) and then sets.

▲ Plaster of Paris is so fine grained that it can be used to make intricate casts. It is, for example, used to make the casts of teeth for people who need to be fitted with dentures.

Bones

Bones are a special form of tissue made by the body cells. Bones are not dead material, but living organs. This is why bones will grow together when broken and why they can make new blood for the body.

A bone contains a mixture of living cells and hard mineral that has been deposited by the cells. The mineral is mainly a compound of calcium (in the form of calcium phosphate).

Bone tissue is always renewing itself, shedding old cells and building new ones. To build new cells, bones store the calcium compounds that reach the body through the blood system.

Calcium and growth

When people are young, they grow fast, and therefore the bone cells need large amounts of calcium so that they can make new bone. This is why it is so vital that the food we eat when we are young contains large amounts of calcium. In later life the body is simply renewing bone, so it needs less calcium.

▲ Because it would not be good for skulls to be made of thick, heavy bone, the calcite of a skull bone is relatively thin. In the case of this beaver skull the ball shape gives strength.

The outer wall of a bone contains plates of calcite. Inside are rods of calcite that carry blood vessels. This is a much softer material, often referred to as bone marrow.

◀▲ Leg and spinal bones are thick because they carry the weight of the body. Nevertheless they are all hollow. This hollow shape, which is similar to a cylinder, is extremely strong. If the bone were solid (like a rod) it would be more easily deformed. A cylinder is also much lighter than a rod.

Teeth

Teeth are a very hard form of calcium carbonate (calcite) specially designed to resist wear. The outer layer of a tooth, called the enamel, has cells arranged in long rods, making them extremely hard and strong.

However, this type of cell is not replaceable in the same way as bones inside the body, which is why teeth do not have the self-repairing properties of other bones and thus must be repaired by dentists.

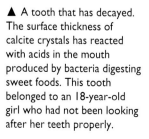

▲ A tooth that has decayed. The surface thickness of calcite crystals has reacted with acids in the mouth produced by bacteria digesting sweet foods. This tooth belonged to an 18-year-old girl who had not been looking after her teeth properly.

Magnesium

Magnesium, like calcium, is a highly reactive element, readily combining with other elements to make compounds.

Magnesium is a silvery metal, not unlike calcium but easier to extract from its ore. It is the most reactive of the metals that can be used in everyday applications (calcium, sodium and potassium are too difficult to produce and are unstable in air, water or both).

The reactivity of magnesium can be easily demonstrated by lighting a taper made of magnesium ribbon. It rapidly bursts into flame giving out a bright white light. (This used to be used for flashlight bulbs; it is still used in signal flares). Similarly, magnesium reacts quickly with dilute hydrochloric acid, giving off hydrogen gas.

▲▶ Magnesium ribbon reacts with the oxygen of the air and develops a dull oxide coating. The end of this sample has been cleaned with emery paper to show the nature of the untarnished metal. The white fragments are of magnesium oxide, formed as the result of burning part of the ribbon (as shown right). The rapid oxidation of the ribbon as it burns is shown by the small pieces of ribbon that are still bright and shiny where the tongs were holding it in the flame.

▲▶ When magnesium ribbon is dropped into dilute hydrochloric acid, a reaction takes place in which hydrogen gas is given off along with a lot of heat.

▲ The most widespread natural use of magnesium is in the chlorophyll in leaves. An ion of magnesium lies at the centre of each chlorophyll molecule.

EQUATION: The reaction of magnesium with an acid

Magnesium + dilute hydrochloric acid ⇨ magnesium chloride + hydrogen gas

$$Mg(s) + 2HCl(aq) \Rightarrow MgCl_2(aq) + H_2(g)$$

Protecting with magnesium

Many metals are subject to corrosion when placed in damp air or damp soil. The most vulnerable of all are iron and steel structures. Small structures can be protected by covering them in a protective coating of, say, paint. However, some iron and steel structures are too big for such treatment. Instead, they are coupled to blocks of metals such as magnesium.

By connecting the metals together in a moist environment, a natural battery is formed. In such a battery, one of the electrodes (the anode) always corrodes, while the other (the cathode) remains undamaged.

The list below shows how magnesium can be used to protect steel. Each metal shown in the list (called a reactivity series) will always protect any metal that comes below it in the series. Those below act as cathodes; those above act as anodes. Thus magnesium will protect exposed iron because it is more reactive, but tin will not.

REACTIVITY SERIES	
Element	Reactivity
potassium	most reactive
sodium	
calcium	
magnesium	
aluminium	
manganese	
chromium	
zinc	
iron	
cadmium	
tin	
lead	
copper	
mercury	
silver	
gold	
platinum	least reactive

Also...

You may be familiar with the protective role of tin as a plating over steel. Here the tin is used as a kind of paint. However, if the tin-plating becomes scratched, you will find the iron corrodes rapidly. This is one reason tin-plating is used less today than in the past.

Magnesium oxide

Just as calcium carbonate (limestone) can be heated to produce calcium oxide (quicklime), so magnesium carbonate (dolomite rock) can be heated to release magnesium oxide.

Magnesium oxide, the most useful compound of magnesium, melts at a very high temperature (2800°C). It is a good conductor of heat, but it conducts electricity poorly. These properties allow it to be used to insulate electric heating elements.

anode: the positive electrode of an electrolysis cell.

cathode: the negative electrode of an electrolysis cell.

corrosion: the *slow* decay of a substance resulting from contact with gases and liquids in the environment. The term is often applied to metals. Rust is the corrosion of iron.

electrode: a conductor that forms one terminal of a cell.

◀▼ The reactivity of magnesium relative to iron is demonstrated here. The liquids in both bottles contain an indicator that turns purple when a reaction takes place. The left-hand bottle contains an iron nail with magnesium wrapped around it and the right-hand bottle contains a similar iron nail with tin wrapped around it.

In the left-hand bottle a reaction takes place at the magnesium strip, causing the formation of magnesium hydroxide, an alkaline substance that makes the indicator turn purple. The iron nail is not corroded because magnesium is more reactive than iron.

By contrast, in the right-hand bottle the tin has not reacted and the iron nail has corroded (rusted). This is because iron is more reactive than tin.

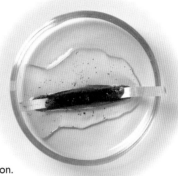

◀▼ When a piece of magnesium ribbon is placed in a dish and some copper sulphate solution is added, the magnesium corrodes very rapidly indeed. This is because magnesium is more reactive than copper. Magnesium is also more reactive than iron, and can be used to protect it.

▼ Oil storage tanks, and other large steel structures, are often protected from corrosion by attaching them to magnesium blocks buried in the ground. The magnesium blocks corrode, while the steel is protected.

Hard water

There are many compounds dissolved in water. Just as invisible bugs in water can cause great health problems, so invisible chemicals can cause a range of problems for people.

Calcium and magnesium compounds are commonly found in water supplies. If water supplies contain more than 120 milligrams of these compounds in each litre of water, the water is described as hard water.

Do you live in a hard-water area?

Most people have a rule of thumb for telling if they live in a hard-water area: if they have trouble getting a lather when in the bath or the shower, then they know they are in a hard-water area.

Most people who live in limestone areas have hard-water supplies, but because water is often transferred for hundreds of kilometres, to match supply and demand, areas far from limestone rocks can also have hard water.

Limescale

Calcium and magnesium carbonates are together usually referred to as limescale. You can sometimes see the thin form of these carbonates build up (precipitate) in many places where hot water is used, such as around the hot water tap in a bathroom, where it gives white smears that thicken to give a dull film.

Carbonates also precipitate quickly on the heating element of a coffee jug or electric kettle, because bicarbonates are not stable in hot water.

▲ You know you are in a hard-water area if scale forms on a boiler or kettle element quickly and looks like this.

Also...

Hard water can be either temporary or permanent. Water containing soluble calcium hydrogen carbonate (calcium bicarbonate) is *temporarily hard* because calcium hydrogen carbonates can be removed by boiling (carbonate precipitates). Water is *permanently hard* if it contains calcium or magnesium salts other than the calcium hydrogen carbonates, as these cannot be removed by boiling. It does not produce scale in kettles but it does make it difficult to obtain a lather with soap.

Removing the scale

A wide range of chemicals has been developed to remove limescale from the kitchen, bathroom, hot water system and boiling elements. They have to remove the lime while being safe on the hands. This means that most mineral acids cannot be used. However, natural organic acids can be used to dissolve the scale. Traditionally acetic acid (vinegar) was used, although modern descaling powders contain citric acid (the same acid as in citrus fruit).

Dangers of scaling

As the carbonate layer thickens, it acts like an artificial stone covering, a sort of cultured stalagmite (see page 14). But a cover of stone cannot conduct heat effectively and so the heating element becomes less efficient. As heat cannot get away from the element, there is also a risk that the element will overheat and burn out.

insoluble: a substance that will not dissolve.

precipitate: tiny solid particles formed as a result of a chemical reaction between two liquids or gases. Limescale is a precipitate from hard water.

▲ You know you are in a hard-water area if soap gives little lather.

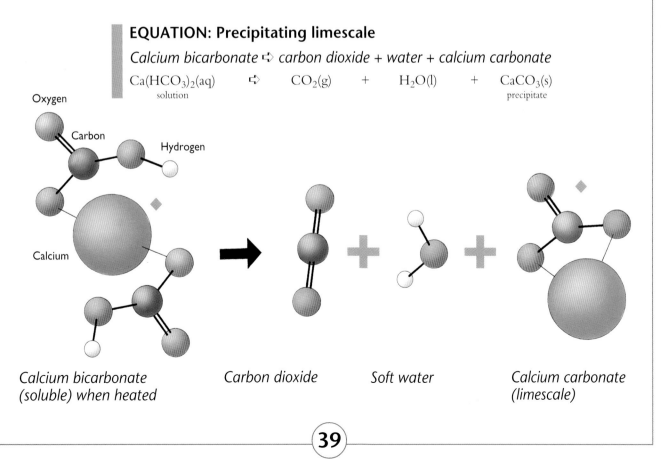

EQUATION: Precipitating limescale

Calcium bicarbonate ⇨ carbon dioxide + water + calcium carbonate

$$Ca(HCO_3)_2(aq) \quad ⇨ \quad CO_2(g) \quad + \quad H_2O(l) \quad + \quad CaCO_3(s)$$

solution   precipitate

Oxygen

Carbon

Hydrogen

Calcium

Calcium bicarbonate (soluble) when heated

Carbon dioxide

Soft water

Calcium carbonate (limescale)

Softening water

Hard water – water containing a large amount of dissolved calcium and magnesium salts – can be difficult to use for washing. It makes the washing up water develop a soapy scum that is hard to remove. This means that it is very difficult to get cutlery and crockery "squeaky clean", making everything less pleasant to look at.

The same fatty scum also builds up out of sight in the drain pipes and can cause blockages that will need expensive maintenance. Hard water also causes hot water systems to scale up and need frequent repair.

Problems of removing hard water with detergents

A more common method of softening water is to put special chemicals in the washing up powder or liquid, so that water is treated as it is used.

A very efficient chemical for this is based on the element phosphorus. This is an element that all plants need for growth. In fact, it is applied to field and gardens as "superphosphate", a kind of fertiliser.

Although the phosphate is good for adding to soils, when used in washing powders phosphate compounds go through sewage works unaltered and enter rivers. As a result, the water organisms such as algae get huge doses of fertiliser, causing them to grow rapidly (a feature known as algal blooms). Later, when the algae die, the bacteria that decompose their remains take most of the oxygen from the water, causing yet more problems. Because of this, phosphate compounds are no longer used to treat hard water. This is why many detergents advertise "phosphate-free" on their packets.

❶ The granules in the water-softener (either clay or a resin) are charged from common salt. The salt breaks down in water, producing sodium ions that stick to a special artificial honeycomb filter.

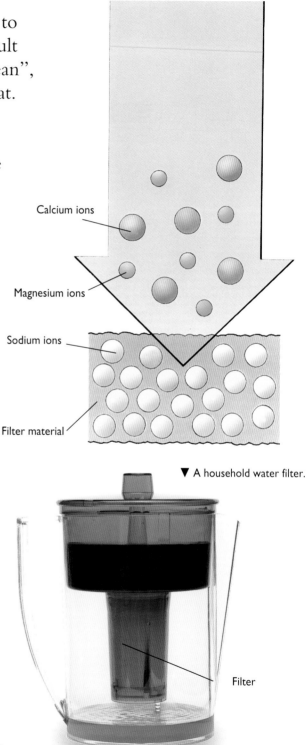

Calcium ions

Magnesium ions

Sodium ions

Filter material

▼ A household water filter.

Filter

How water-softeners work

Although most people are prepared to put up with hard-water problems, it is possible to use a chemical means to soften the water. Water-softening is a method of taking away the substances that cause the water to be hard.

One commonly used method is to pass the water entering the house through a tank containing a filter. Modern filters use a special clay mineral or an artificial resin.

The filter absorbs on to its surface material such as common salt, which will react with the calcium and magnesium compounds.

ion: an atom, or group of atoms, that has gained or lost one or more electrons and so developed an electrical charge.

resin: natural or synthetic polymers that can be moulded into solid objects or spun into thread.

2 The hard water washes over the filter and the sodium and calcium ions exchange until they are in balance. This takes some of the calcium and magnesium ions out of the water, making it less hard.

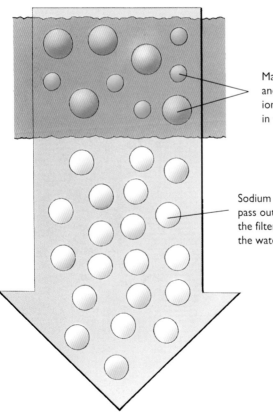

Magnesium and calcium ions are left in the filter.

Sodium ions pass out of the filter into the water.

▲ This water-softener passes the incoming hard water over artificial resin spheres.

3 From time to time the filter is recharged by flushing it with salt. This exchanges calcium and magnesium ions for sodium ions.

EQUATION: Water-softening
Hard water (water containing calcium and magnesium compounds in solution)… exchanges for the sodium in salt… leaving calcium and magnesium behind.

Also...

A compound may be bonded together by strong electrical forces. When such a compound dissolves, it splits up, with each of the parts retaining its own electrical charge.

These charged particles are called ions. Calcium ions and magnesium ions have a positive charge; bicarbonate ions have a negative charge. When they re-form into a compound, it will be a neutral substance called a salt.

Water-softening treatments rely on the way that ions can be swapped in and out of a solution.

Antacids

The body makes strong acids to help to digest food. Most of these acids are produced in the lining of the stomach.

The acids are vital and usually the food neutralises the acids exactly, so that no discomfort is produced. But under some circumstances, especially if we eat a large amount of food and do not chew it properly, the digestive system does not balance.

Neutralising

One way to counteract the occasional problem of indigestion is to use a substance, such as calcium carbonate, that reacts with an acid. "Antacids" are normally taken in powder form, as a tablet, or sometimes as a suspension in water. Calcium bicarbonate is also used because it is soluble and can be mixed more easily with water.

Magnesium hydroxide can be mixed in water and made into a suspension called "Milk of Magnesia". Because it is only a suspension, over time the magnesium hydroxide settles to the bottom of the bottle. This is why it is important to shake the bottle before use.

Magnesium hydroxide has an advantage over calcium carbonate. When it reacts with acids, it does not produce carbon dioxide, so there is no embarrassing burping after taking milk of magnesia, as there might be after taking calcium bicarbonate!

▲ Magnesium hydroxide, Milk of Magnesia, is a traditional antacid. Notice that it is a mixture (a suspension of magnesium hydroxide in water). Over time the suspension settles out. The precipitate shows clearly at the bottom of this bottle.

> ## EQUATION: Neutralising stomach acid with a hydroxide
> *Hydrochloric acid + magnesium hydroxide ⇨ magnesium chloride + water*
> $$2HCl(aq) + Mg(OH)_2(aq) \Rightarrow MgCl_2(aq) + 2H_2O(l)$$

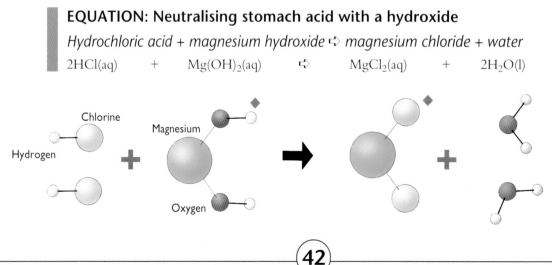

Chlorine

Magnesium

Hydrogen

Oxygen

The gas that tells of success

The action of a carbonate antacid is to produce a gas. When this builds up in the digestive system it is released quite uncontrollably as a "burp".

However, the gas that occurs in some fizzing antacid tablets is formed in the glass. This is because the manufacturers put citric acid powder in with the calcium carbonate. When the tablet is put in water, the acid and carbonate react to release carbon dioxide and give an attractive impression of action even before the antacid is digested.

acid: compounds containing hydrogen which can attack and dissolve many substances.

neutralisation: the reaction of acids and bases to produce a salt and water. The reaction causes hydrogen from the acid and hydroxide from the base to be changed to water. For example, hydrochloric acid reacts with sodium hydroxide to form common salt and water. The term is more generally used for any reaction where the pH changes towards 7.0, which is the pH of a neutral solution.

Carbon dioxide gas in burp

◄► Indigestion tablets may contain calcium or magnesium carbonate or bicarbonate. Here you can see some tablets reacting vigorously with dilute hydrochloric acid just as takes place in your stomach. The bubbles are carbon dioxide gas.

Why too much acid is produced

The digestive system is triggered into releasing acids to match the volume of food we eat. So if we eat a lot, a great deal of acid is released from the linings of the digestive system.

Acids work on the surfaces of food, so the better it is chewed, the faster the acids can get to work. If food is eaten in large lumps (i.e. it is bolted down), the acid will not be able to get to it quickly. As a result the amount of acid produced can be more than is needed for digestion. The excess acid then starts to attack the linings of the digestive system, causing the pains that are known as indigestion and heartburn. Bacteria working in this very acid environment can even cause ulcers.

Acid is released from stomach lining.

Calcium carbonate reacts with acids.

Digestive juices in stomach are broken down by acids.

EQUATION: Neutralising stomach acid with a carbonate

Hydrochloric acid + calcium carbonate ⇨ calcium chloride + water + carbon dioxide gas

$$2HCl(aq) \quad + \quad CaCO_3(s) \quad ⇨ \quad CaCl_2(aq) \quad + \quad 2H_2O(l) \quad + \quad CO_2(g)$$

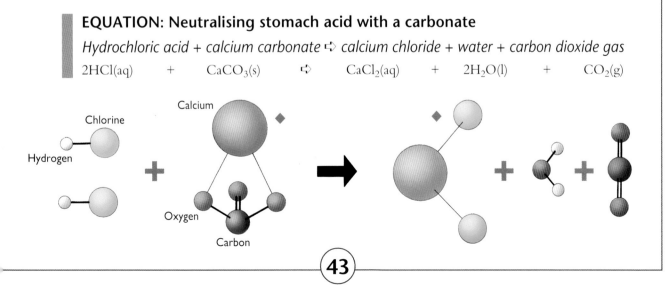

Hydrogen
Chlorine
Calcium
Oxygen
Carbon

Key facts about...

Calcium

A soft, silvery-coloured metal, chemical symbol Ca

Essential for plant and animal growth

Fifth most plentiful element at the surface of the Earth

Calcium carbonate is insoluble in water

Forms the minerals that make limestone rocks

Calcium bicarbonate dissolved in cold water causes "temporary hardness"

Is part of the bones and shells of all animals

Found most commonly as calcium carbonate, calcite

Makes up over 3% of the Earth's crust

Atomic number 20, atomic weight about 40

SHELL DIAGRAMS

The shell diagrams on these two pages are representations of an atom of each element. The total number of electrons is shown in the relevant orbitals, or shells, around the central nucleus.

Electron shell

Electron

Nucleus containing protons and neutrons (called nucleons)

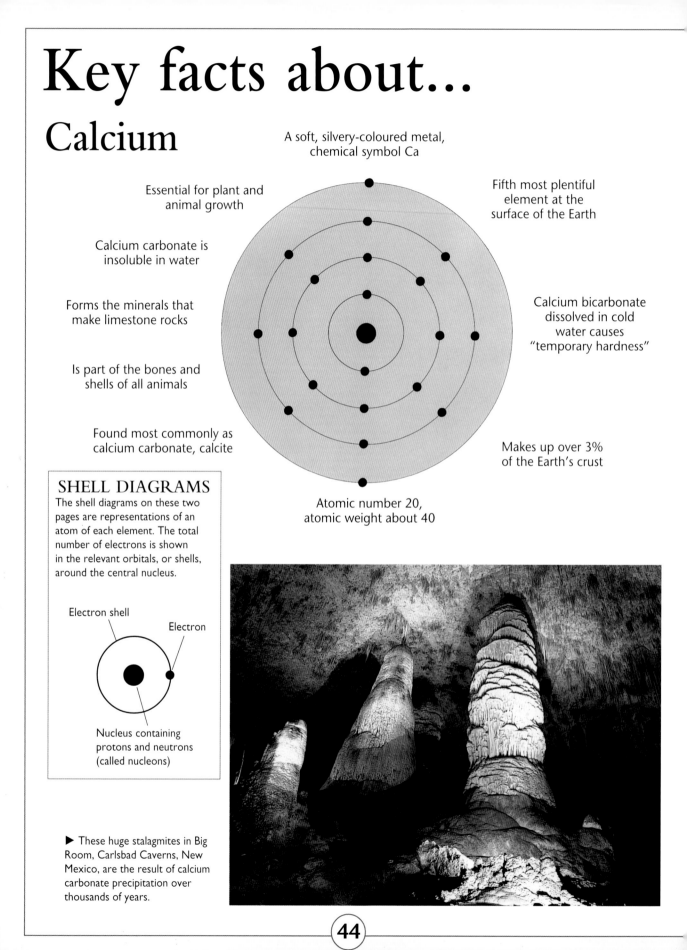

▶ These huge stalagmites in Big Room, Carlsbad Caverns, New Mexico, are the result of calcium carbonate precipitation over thousands of years.

44

Magnesium

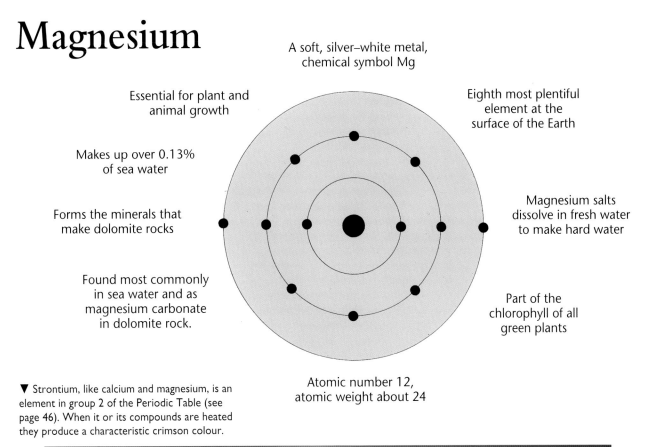

A soft, silver–white metal, chemical symbol Mg

Essential for plant and animal growth

Eighth most plentiful element at the surface of the Earth

Makes up over 0.13% of sea water

Forms the minerals that make dolomite rocks

Magnesium salts dissolve in fresh water to make hard water

Found most commonly in sea water and as magnesium carbonate in dolomite rock.

Part of the chlorophyll of all green plants

Atomic number 12, atomic weight about 24

▼ Strontium, like calcium and magnesium, is an element in group 2 of the Periodic Table (see page 46). When it or its compounds are heated they produce a characteristic crimson colour.

The Periodic Table

The Periodic Table sets out the relationships among the elements of the Universe. According to the Periodic Table, certain elements fall into groups. The pattern of these groups has, in the past, allowed scientists to predict elements that had not at that time been discovered. It can still be used today to predict the properties of unfamiliar elements.

The Periodic Table was first described by a Russian teacher, Dmitry Ivanovich Mendeleev, between 1869 and 1870. He was interested in writing a chemistry textbook, and wanted to show his students that there were certain patterns in the elements that had been discovered. So he set out the elements (of which there were 57 at the time) according to their known properties. On the assumption that there was pattern to the elements, he left blank spaces where elements seemed to be missing. Using this first version of the Periodic Table, he was able to predict in detail the chemical and physical properties of elements that had not yet been discovered. Other scientists began to look for the missing elements, and they soon found them.

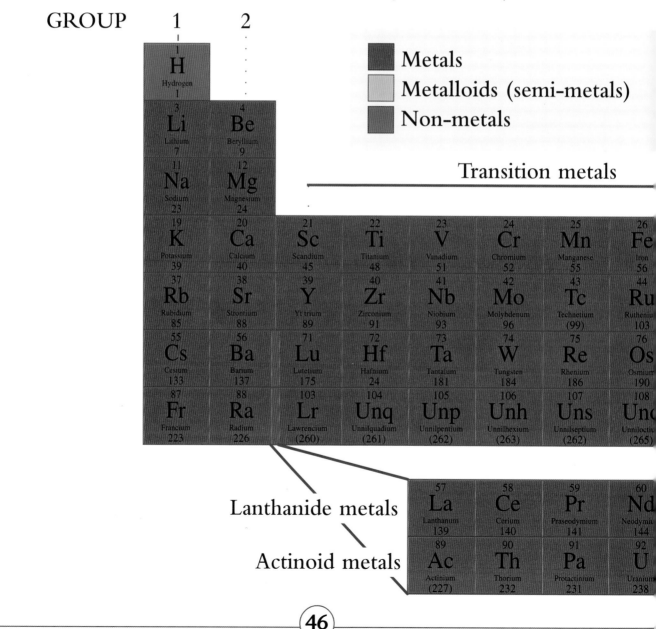

Hydrogen did not seem to fit into the table, so he placed it in a box on its own. Otherwise the elements were all placed horizontally. When an element was reached with properties similar to the first one in the top row, a second row was started. By following this rule, similarities among the elements can be found by reading up and down. By reading across the rows, the elements progressively increase their atomic number. This number indicates the number of positively charged particles (protons) in the nucleus of each atom. This is also the number of negatively charged particles (electrons) in the atom.

The chemical properties of an element depend on the number of electrons in the outermost shell.

Atoms can form compounds by sharing electrons in their outermost shells. This explains why atoms with a full set of electrons (like helium, an inert gas) are unreactive, whereas atoms with an incomplete electron shell (such as chlorine) are very reactive. Elements can also combine by the complete transfer of electrons from metals to non-metals and the compounds formed contain ions.

Radioactive elements lose particles from their nucleus and electrons from their surrounding shells. As a result their atomic number changes and they become new elements.

Atomic (proton) number

13 — **Symbol**

Al

Aluminium — **Name**

27 — **Approximate atomic weight**

3	4	5	6	7	0
					2 **He** Helium 4
5 **B** Boron 11	6 **C** Carbon 12	7 **N** Nitrogen 14	8 **O** Oxygen 16	9 **F** Fluorine 19	10 **Ne** Neon 20
13 **Al** Aluminium 27	14 **Si** Silicon 28	15 **P** Phosphorus 31	16 **S** Sulphur 32	17 **Cl** Chlorine 35	18 **Ar** Argon 40

27 **Co** Cobalt 59	28 **Ni** Nickel 59	29 **Cu** Copper 64	30 **Zn** Zinc 65	31 **Ga** Gallium 70	32 **Ge** Germanium 73	33 **As** Arsenic 75	34 **Se** Selenium 79	35 **Br** Bromine 80	36 **Kr** Krypton 84
45 **Rh** Rhodium 103	46 **Pd** Palladium 106	47 **Ag** Silver 108	48 **Cd** Cadmium 112	49 **In** Indium 115	50 **Sn** Tin 119	51 **Sb** Antimony 122	52 **Te** Tellurium 128	53 **I** Iodine 127	54 **Xe** Xenon 131
77 **Ir** Iridium 192	78 **Pt** Platinum 195	79 **Au** Gold 197	80 **Hg** Mercury 201	81 **Tl** Thallium 204	82 **Pb** Lead 207	83 **Bi** Bismuth 209	84 **Po** Polonium (209)	85 **At** Astatine (210)	86 **Rn** Radon (222)
109 **Une** Unnilennium (266)									

61 **Pm** Promethium (145)	62 **Sm** Samarium 150	63 **Eu** Europium 152	64 **Gd** Gadolinium 157	65 **Tb** Terbium 159	66 **Dy** Dysprosium 163	67 **Ho** Holmium 165	68 **Er** Erbium 167	69 **Tm** Thulium 169	70 **Yb** Ytterbium 173
93 **Np** Neptunium (237)	94 **Pu** Plutonium (244)	95 **Am** Americium (243)	96 **Cm** Curium (247)	97 **Bk** Berkelium (247)	98 **Cf** Californium (251)	99 **Es** Einsteinium (252)	100 **Fm** Fermium (257)	101 **Md** Mendelevium (258)	102 **No** Nobelium (259)

Understanding equations

As you read through this book, you will notice that many pages contain equations using symbols. If you are not familiar with these symbols, read this page. Symbols make it easy for chemists to write out the reactions that are occurring in a way that allows a better understanding of the processes involved.

Symbols for the elements

The basis of the modern use of symbols for elements dates back to the 19th century. At this time a shorthand was developed using the first letter of the element wherever possible. Thus "O" stands for oxygen, "H" stands for hydrogen

and so on. However, if we were to use only the first letter, then there could be some confusion. For example, nitrogen and nickel would both use the symbols N. To overcome this problem, many elements are symbolised using the first two letters of their full name, and the second letter is lowercase. Thus although nitrogen is N, nickel becomes Ni. Not all symbols come from the English name; many use the Latin name instead. This is why, for example, gold is not G but Au (for the Latin *aurum*) and sodium has the symbol Na, from the Latin *natrium*.

Compounds of elements are made by combining letters. Thus the molecule carbon

Written and symbolic equations

In this book, important chemical equations are briefly stated in words (these are called word equations), and are then shown in their symbolic form along with the states.

What reaction the equation illustrates

Word equation

Symbol equation

Sometimes you will find additional descriptions below the symbolic equation.

EQUATION: The formation of calcium hydroxide

Calcium oxide + water ⇨ calcium hydroxide

$$CaO(s) \quad + \quad H_2O(l) \quad \overset{\Rightarrow}{\underset{\text{heated}}{}} \quad Ca(OH)_2(aq)$$

Symbol showing the state: s is for solid, l is for liquid, g is for gas and aq is for aqueous.

Diagrams

Some of the equations are shown as graphic representations.

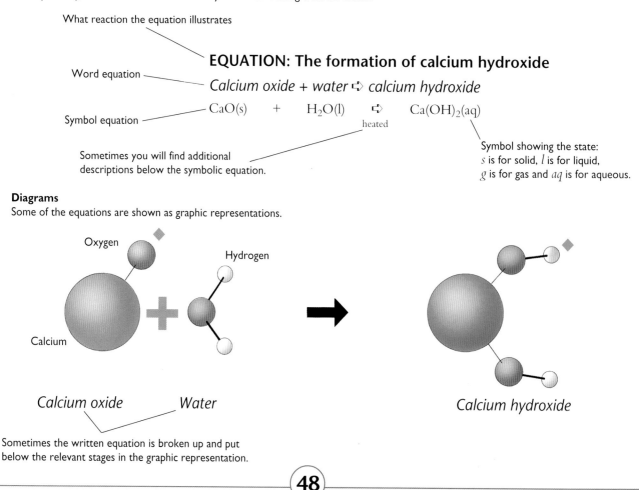

Oxygen

Hydrogen

Calcium

Calcium oxide   Water

Calcium hydroxide

Sometimes the written equation is broken up and put below the relevant stages in the graphic representation.

monoxide is CO. By using lowercase letters for the second letter of an element, it is possible to show that cobalt, symbol Co, is not the same as the molecule carbon monoxide, CO.

However, the letters can be made to do much more than this. In many molecules, atoms combine in unequal numbers. So, for example, carbon dioxide has one atom of carbon for every two of oxygen. This is shown by using the number 2 beside the oxygen, and the symbol becomes CO_2.

In practice, some groups of atoms combine as a unit with other substances. Thus, for example, calcium bicarbonate (one of the compounds used in some antacid pills) is written $Ca(HCO_3)_2$. This shows that the part of the substance inside the brackets reacts as a unit and the "2" outside the brackets shows the presence of two such units.

Some substances attract water molecules to themselves. To show this a dot is used. Thus the blue form of copper sulphate is written $CuSO_4.5H_2O$. In this case five molecules of water attract to one of copper sulphate.

Atoms and ions
Each sphere represents a particle of an element. A particle can be an atom or an ion. Each atom or ion is associated with other atoms or ions through bonds – forces of attraction. The size of the particles and the nature of the bonds can be extremely important in determining the nature of the reaction or the properties of the compound.

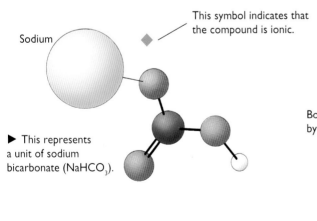

Sodium

This symbol indicates that the compound is ionic.

▶ This represents a unit of sodium bicarbonate ($NaHCO_3$).

The term "unit" is sometimes used to simplify the representation of a combination of ions.

When you see the dot, you know that this water can be driven off by heating; it is part of the crystal structure.

In a reaction substances change by rearranging the combinations of atoms. The way they change is shown by using the chemical symbols, placing those that will react (the starting materials, or reactants) on the left and the products of the reaction on the right. Between the two, chemists use an arrow to show which way the reaction is occurring.

It is possible to describe a reaction in words. This gives word equations, which are given throughout this book. However, it is easier to understand what is happening by using an equation containing symbols. These are also given in many places. They are not given when the equations are very complex.

In any equation both sides balance; that is, there must be an equal number of like atoms on both sides of the arrow. When you try to write down reactions, you, too, must balance your equation; you cannot have a few atoms left over at the end!

The symbols in brackets are abbreviations for the physical state of each substance taking part, so that (*s*) is used for solid, (*l*) for liquid, (*g*) for gas and (*aq*) for an aqueous solution, that is, a solution of a substance dissolved in water.

Chemical symbols, equations and diagrams
The arrangement of any molecule or compound can be shown in one of the two ways shown below, depending on which gives the clearer picture. The left-hand diagram is called a ball-and-stick diagram because it uses rods and spheres to show the structure of the material. This example shows water, H_2O. There are two hydrogen atoms and one oxygen atom.

Bond shown by "stick"

Colours too
The colours of each of the particles help differentiate the elements involved. The diagram can then be matched to the written and symbolic equation given with the diagram. In the case above, oxygen is red and hydrogen is grey.

Glossary of technical terms

absorb: to soak up a substance. Compare to adsorb.

acetone: a petroleum-based solvent.

acid: compounds containing hydrogen which can attack and dissolve many substances. Acids are described as weak or strong, dilute or concentrated, mineral or organic.

acidity: a general term for the strength of an acid in a solution.

acid rain: rain that is contaminated by acid gases such as sulphur dioxide and nitrogen oxides released by pollution.

adsorb/adsorption: to "collect" gas molecules or other particles on to the *surface* of a substance. They are not chemically combined and can be removed. (The process is called "adsorption".) Compare to absorb.

alchemy: the traditional "art" of working with chemicals that prevailed through the Middle Ages. One of the main challenges of alchemy was to make gold from lead. Alchemy faded away as scientific chemistry was developed in the 17th century.

alkali: a base in solution.

alkaline: the opposite of acidic. Alkalis are bases that dissolve, and alkaline materials are called basic materials. Solutions of alkalis have a pH greater than 7.0 because they contain relatively few hydrogen ions.

alloy: a mixture of a metal and various other elements.

alpha particle: a stable combination of two protons and two neutrons, which is ejected from the nucleus of a radioactive atom as it decays. An alpha particle is also the nucleus of the atom of helium. If it captures two electrons it can become a neutral helium atom.

amalgam: a liquid alloy of mercury with another metal.

amino acid: amino acids are organic compounds that are the building blocks for the proteins in the body.

amorphous: a solid in which the atoms are not arranged regularly (i.e. "glassy"). Compare with crystalline.

amphoteric: a metal that will react with both acids and alkalis.

anhydrous: a substance from which water has been removed by heating. Many hydrated salts are crystalline. When they are heated and the water is driven off, the material changes to an anhydrous powder.

anion: a negatively charged atom or group of atoms.

anode: the negative terminal of a battery or the positive electrode of an electrolysis cell.

anodising: a process that uses the effect of electrolysis to make a surface corrosion-resistant.

antacid: a common name for any compound that reacts with stomach acid to neutralise it.

antioxidant: a substance that prevents oxidation of some other substance.

aqueous: a solid dissolved in water. Usually used as "aqueous solution".

atom: the smallest particle of an element.

atomic number: the number of electrons or the number of protons in an atom.

atomised: broken up into a very fine mist. The term is used in connection with sprays and engine fuel systems.

aurora: the "northern lights" and "southern lights" that show as coloured bands of light in the night sky at high latitudes. They are associated with the way cosmic rays interact with oxygen and nitrogen in the air.

basalt: an igneous rock with a low proportion of silica (usually below 55%). It has microscopically small crystals.

base: a compound that may be soapy to the touch and that can react with an acid in water to form a salt and water.

battery: a series of electrochemical cells.

bauxite: an ore of aluminium, of which about half is aluminium oxide.

becquerel: a unit of radiation equal to one nuclear disintegration per second.

beta particle: a form of radiation in which electrons are emitted from an atom as the nucleus breaks down.

bleach: a substance that removes stains from materials either by oxidising or reducing the staining compound.

boiling point: the temperature at which a liquid boils, changing from a liquid to a gas.

bond: chemical bonding is either a transfer or sharing of electrons by two or more atoms. There are a number of types of chemical bond, some very strong (such as covalent bonds), others weak (such as hydrogen bonds). Chemical bonds form because the linked molecule is more stable than the unlinked atoms from which it formed. For example, the hydrogen molecule (H_2) is more stable than single atoms of hydrogen, which is why hydrogen gas is always found as molecules of two hydrogen atoms.

brass: a metal alloy principally of copper and zinc.

brazing: a form of soldering, in which brass is used as the joining metal.

brine: a solution of salt (sodium chloride) in water.

bronze: an alloy principally of copper and tin.

buffer: a chemistry term meaning a mixture of substances in solution that resists a change in the acidity or alkalinity of the solution.

capillary action: the tendency of a liquid to be sucked into small spaces, such as between objects and through narrow-pore tubes. The force to do this comes from surface tension.

catalyst: a substance that speeds up a chemical reaction but itself remains unaltered at the end of the reaction.

cathode: the positive terminal of a battery or the negative electrode of an electrolysis cell.

cathodic protection: the technique of making the object that is to be protected from corrosion into the cathode of a cell. For example, a material, such as steel, is protected by coupling it with a more reactive metal, such as magnesium. Steel forms the cathode and magnesium the anode. Zinc protects steel in the same way.

cation: a positively charged atom or group of atoms.

caustic: a substance that can cause burns if it touches the skin.

cell: a vessel containing two electrodes and an electrolyte that can act as an electrical conductor.

ceramic: a material based on clay minerals, which has been heated so that it has chemically hardened.

chalk: a pure form of calcium carbonate made of the crushed bodies of microscopic sea creatures, such as plankton and algae.

change of state: a change between one of the three states of matter, solid, liquid and gas.

chlorination: adding chlorine to a substance.

cladding: a surface sheet of material designed to protect other materials from corrosion.

clay: a microscopically small plate-like mineral that makes up the bulk of many soils. It has a sticky feel when wet.

combustion: the special case of oxidisation of a substance where a considerable amount of heat and usually light are given out. Combustion is often referred to as "burning".

compound: a chemical consisting of two or more elements chemically bonded together. Calcium atoms can combine with carbon atoms and oxygen atoms to make calcium carbonate, a compound of all three atoms.

condensation nuclei: microscopic particles of dust, salt and other materials suspended in the air, which attract water molecules.

conduction: (i) the exchange of heat (heat conduction) by contact with another object or (ii) allowing the flow of electrons (electrical conduction).

convection: the exchange of heat energy with the surroundings produced by the flow of a fluid due to being heated or cooled.

corrosion: the *slow* decay of a substance resulting from contact with gases and liquids in the environment. The term is often applied to metals. Rust is the corrosion of iron.

corrosive: a substance, either an acid or an alkali, that *rapidly* attacks a wide range of other substances.

cosmic rays: particles that fly through space and bombard all atoms on the Earth's surface. When they interact with the atmosphere they produce showers of secondary particles.

covalent bond: the most common form of strong chemical bonding, which occurs when two atoms *share* electrons.

cracking: breaking down complex molecules into simpler components. It is a term particularly used in oil refining.

crude oil: a chemical mixture of petroleum liquids. Crude oil forms the raw material for an oil refinery.

crystal: a substance that has grown freely so that it can develop external faces. Compare with crystalline, where the atoms are not free to form individual crystals and amorphous where the atoms are arranged irregularly.

crystalline: the organisation of atoms into a rigid "honeycomb-like" pattern without distinct crystal faces.

crystal systems: seven patterns or systems into which all of the world's crystals can be grouped. They are: cubic, hexagonal, rhombohedral, tetragonal, orthorhombic, monoclinic and triclinic.

cubic crystal system: groupings of crystals that look like cubes.

curie: a unit of radiation. The amount of radiation emitted by 1 g of radium each second. (The curie is equal to 37 billion becquerels.)

current: an electric current is produced by a flow of electrons through a conducting solid or ions through a conducting liquid.

decay (radioactive decay): the way that a radioactive element changes into another element because of loss of mass through radiation. For example uranium decays (changes) to lead.

decompose: to break down a substance (for example by heat or with the aid of a catalyst) into simpler components. In such a chemical reaction only one substance is involved.

dehydration: the removal of water from a substance by heating it, placing it in a dry atmosphere, or through the action of a drying agent.

density: the mass per unit volume (e.g. g/cc).

desertification: a process whereby a soil is allowed to become degraded to a state in which crops can no longer grow, i.e. desert-like. Chemical desertification is usually the result of contamination with halides because of poor irrigation practices.

detergent: a petroleum-based chemical that removes dirt.

diaphragm: a semipermeable membrane – a kind of ultra-fine mesh filter – that will allow only small ions to pass through. It is used in the electrolysis of brine.

diffusion: the slow mixing of one substance with another until the two substances are evenly mixed.

digestive tract: the system of the body that forms the pathway for food and its waste products. It begins at the mouth and includes the stomach and the intestines.

dilute acid: an acid whose concentration has been reduced by a large proportion of water.

diode: a semiconducting device that allows an electric current to flow in only one direction.

disinfectant: a chemical that kills bacteria and other microorganisms.

dissociate: to break apart. In the case of acids it means to break up forming hydrogen ions. This is an example of ionisation. Strong acids dissociate completely. Weak acids are not completely ionised and a solution of a weak acid has a relatively low concentration of hydrogen ions.

dissolve: to break down a substance in a solution without a resultant reaction.

distillation: the process of separating mixtures by condensing the vapours through cooling.

doping: adding metal atoms to a region of silicon to make it semiconducting.

dye: a coloured substance that will stick to another substance, so that both appear coloured.

electrode: a conductor that forms one terminal of a cell.

electrolysis: an electrical–chemical process that uses an electric current to cause the break up of a compound and the movement of metal ions in a solution. The process happens in many natural situations (as for example in rusting) and is also commonly used in industry for purifying (refining) metals or for plating metal objects with a fine, even metal coating.

electrolyte: a solution that conducts electricity.

electron: a tiny, negatively charged particle that is part of an atom. The flow of electrons through a solid material such as a wire produces an electric current.

electroplating: depositing a thin layer of a metal onto the surface of another substance using electrolysis.

element: a substance that cannot be decomposed into simpler substances by chemical means

emulsion: tiny droplets of one substance dispersed in another. A common oil in water emulsion is milk. The tiny droplets in an emulsion tend to come together, so another stabilising substance is often needed to wrap the particles of grease and oil in a stable coat. Soaps and detergents are such agents. Photographic film is an example of a solid emulsion.

endothermic reaction: a reaction that takes heat from the surroundings. The reaction of carbon monoxide with a metal oxide is an example.

enzyme: organic catalysts in the form of proteins in the body that speed up chemical reactions. Every living cell contains hundreds of enzymes, which ensure that the processes of life continue. Should enzymes be made inoperative, such as through mercury poisoning, then death follows.

ester: organic compounds, formed by the reaction of an alcohol with an acid, which often have a fruity taste.

evaporation: the change of state of a liquid to a gas. Evaporation happens below the boiling point and is used as a method of separating out the materials in a solution.

exothermic reaction: a reaction that gives heat to the surroundings. Many oxidation reactions, for example, give out heat.

explosive: a substance which, when a shock is applied to it, decomposes very rapidly, releasing a very large amount of heat and creating a large volume of gases as a shock wave.

extrusion: forming a shape by pushing it through a die. For example, toothpaste is extruded through the cap (die) of the toothpaste tube.

fallout: radioactive particles that reach the ground from radioactive materials in the atmosphere.

fat: semi-solid energy-rich compounds derived from plants or animals and which are made of carbon, hydrogen and oxygen. Scientists call these esters.

feldspar: a mineral consisting of sheets of aluminium silicate. This is the mineral from which the clay in soils is made.

fertile: able to provide the nutrients needed for unrestricted plant growth.

filtration: the separation of a liquid from a solid using a membrane with small holes.

fission: the breakdown of the structure of an atom, popularly called "splitting the atom" because the atom is split into approximately two other nuclei. This is different from, for example, the small change that happens when radioactivity is emitted.

fixation of nitrogen: the processes that natural organisms, such as bacteria, use to turn the nitrogen of the air into ammonium compounds.

fixing: making solid and liquid nitrogen-containing compounds from nitrogen gas. The compounds that are formed can be used as fertilisers.

fluid: able to flow; either a liquid or a gas.

fluorescent: a substance that gives out visible light when struck by invisible waves such as ultraviolet rays.

flux: a material used to make it easier for a liquid to flow. A flux dissolves metal oxides and so prevents a metal from oxidising while being heated.

foam: a substance that is sufficiently gelatinous to be able to contain bubbles of gas. The gas bulks up the substance, making it behave as though it were semi-rigid.

fossil fuels: hydrocarbon compounds that have been formed from buried plant and animal remains. High pressures and temperatures lasting over millions of years are required. The fossil fuels are coal, oil and natural gas.

fraction: a group of similar components of a mixture. In the petroleum industry the light fractions of crude oil are those with the smallest molecules, while the medium and heavy fractions have larger molecules.

free radical: a very reactive atom or group with a "spare" electron.

freezing point: the temperature at which a substance changes from a liquid to a solid. It is the same temperature as the melting point.

fuel: a concentrated form of chemical energy. The main sources of fuels (called fossil fuels because they were formed by geological processes) are coal, crude oil and natural gas. Products include methane, propane and gasoline. The fuel for stars and space vehicles is hydrogen.

fuel rods: rods of uranium or other radioactive material used as a fuel in nuclear power stations.

fuming: an unstable liquid that gives off a gas. Very concentrated acid solutions are often fuming solutions.

fungicide: any chemical that is designed to kill fungi and control the spread of fungal spores.

fusion: combining atoms to form a heavier atom.

galvanising: applying a thin zinc coating to protect another metal.

gamma rays: waves of radiation produced as the nucleus of a radioactive element rearranges itself into a tighter cluster of protons and neutrons. Gamma rays carry enough energy to damage living cells.

gangue: the unwanted material in an ore.

gas: a form of matter in which the molecules form no definite shape and are free to move about to fill any vessel they are put in.

gelatinous: a term meaning made with water. Because a gelatinous precipitate is mostly water, it is of a similar density to water and will float or lie suspended in the liquid.

gelling agent: a semi-solid jelly-like substance.

gemstone: a wide range of minerals valued by people, both as crystals (such as emerald) and as decorative stones (such as agate). There is no single chemical formula for a gemstone.

glass: a transparent silicate without any crystal growth. It has a glassy lustre and breaks with a curved fracture. Note that some minerals have all these features and are therefore natural glasses. Household glass is a synthetic silicate.

glucose: the most common of the natural sugars. It occurs as the polymer known as cellulose, the fibre in plants. Starch is also a form of glucose. The breakdown of glucose provides the energy that animals need for life.

granite: an igneous rock with a high proportion of silica (usually over 65%). It has well-developed large crystals. The largest pink, grey or white crystals are feldspar.

Greenhouse Effect: an increase of the global air temperature as a result of heat released from burning fossil fuels being absorbed by carbon dioxide in the atmosphere.

gypsum: the name for calcium sulphate. It is commonly found as Plaster of Paris and wallboards.

half-life: the time it takes for the radiation coming from a sample of a radioactive element to decrease by half.

halide: a salt of one of the halogens (fluorine, chlorine, bromine and iodine).

halite: the mineral made of sodium chloride.

halogen: one of a group of elements including chlorine, bromine, iodine and fluorine.

heat-producing: see exothermic reaction.

high explosive: a form of explosive that will only work when it receives a shock from another explosive. High explosives are much more powerful than ordinary explosives. Gunpowder is not a high explosive.

hydrate: a solid compound in crystalline form that contains molecular water. Hydrates commonly form when a solution of a soluble salt is evaporated. The water that forms part of a hydrate crystal is known as the "water of crystallization". It can usually be removed by heating, leaving an anhydrous salt.

hydration: the absorption of water by a substance. Hydrated materials are not "wet" but remain firm, apparently dry, solids. In some cases, hydration makes the substance change colour, in many other cases there is no colour change, simply a change in volume.

hydrocarbon: a compound in which only hydrogen and carbon atoms are present. Most fuels are hydrocarbons, as is the simple plastic polyethene (known as polythene).

hydrogen bond: a type of attractive force that holds one molecule to another. It is one of the weaker forms of intermolecular attractive force.

hydrothermal: a process in which hot water is involved. It is usually used in the context of rock formation because hot water and other fluids sent outwards from liquid magmas are important carriers of metals and the minerals that form gemstones.

igneous rock: a rock that has solidified from molten rock, either volcanic lava on the Earth's surface or magma deep underground. In either case the rock develops a network of interlocking crystals.

incendiary: a substance designed to cause burning.

indicator: a substance or mixture of substances that change colour with acidity or alkalinity.

inert: nonreactive.

infra-red radiation: a form of light radiation where the wavelength of the waves is slightly longer than visible light. Most heat radiation is in the infra-red band.

insoluble: a substance that will not dissolve.

ion: an atom, or group of atoms, that has gained or lost one or more electrons and so developed an electrical charge. Ions behave differently from electrically neutral atoms and molecules. They can move in an electric field, and they can also bind strongly to solvent molecules such as water. Positively charged ions are called cations; negatively charged ions are called anions. Ions carry electrical current through solutions.

ionic bond: the form of bonding that occurs between two ions when the ions have opposite charges. Sodium cations bond with chloride anions to form common salt (NaCl) when a salty solution is evaporated. Ionic bonds are strong bonds except in the presence of a solvent.

ionise: to break up neutral molecules into oppositely charged ions or to convert atoms into ions by the loss of electrons.

ionisation: a process that creates ions.

irrigation: the application of water to fields to help plants grow during times when natural rainfall is sparse.

isotope: atoms that have the same number of protons in their nucleus, but which have different masses; for example, carbon-12 and carbon-14.

latent heat: the amount of heat that is absorbed or released during the process of changing state between gas, liquid or solid. For example, heat is absorbed when a substance melts and it is released again when the substance solidifies.

latex: (the Latin word for "liquid") a suspension of small polymer particles in water. The rubber that flows from a rubber tree is a natural latex. Some synthetic polymers are made as latexes, allowing polymerisation to take place in water.

lava: the material that flows from a volcano.

limestone: a form of calcium carbonate rock that is often formed of lime mud. Most limestones are light grey and have abundant fossils.

liquid: a form of matter that has a fixed volume but no fixed shape.

lode: a deposit in which a number of veins of a metal found close together.

lustre: the shininess of a substance.

magma: the molten rock that forms a balloon-shaped chamber in the rock below a volcano. It is fed by rock moving upwards from below the crust.

marble: a form of limestone that has been "baked" while deep inside mountains. This has caused the limestone to melt and reform into small interlocking crystals, making marble harder than limestone.

mass: the amount of matter in an object. In everyday use, the word weight is often used to mean mass.

melting point: the temperature at which a substance changes state from a solid to a liquid. It is the same as freezing point.

membrane: a thin flexible sheet. A semipermeable membrane has microscopic holes of a size that will selectively allow some ions and molecules to pass through but hold others back. It thus acts as a kind of sieve.

meniscus: the curved surface of a liquid that forms when it rises in a small bore, or capillary tube. The meniscus is convex (bulges upwards) for mercury and is concave (sags downwards) for water.

metal: a substance with a lustre, the ability to conduct heat and electricity and which is not brittle.

metallic bonding: a kind of bonding in which atoms reside in a "sea" of mobile electrons. This type of bonding allows metals to be good conductors and means that they are not brittle

metamorphic rock: formed either from igneous or sedimentary rocks, by heat and or pressure. Metamorphic rocks form deep inside mountains during periods of mountain building. They result from the remelting of rocks during which process crystals are able to grow. Metamorphic rocks often show signs of banding and partial melting.

micronutrient: an element that the body requires in small amounts. Another term is trace element.

mineral: a solid substance made of just one element or chemical compound. Calcite is a mineral because it consists only of calcium carbonate, halite is a mineral because it contains only sodium chloride, quartz is a mineral because it consists of only silicon dioxide.

mineral acid: an acid that does not contain carbon and that attacks minerals. Hydrochloric, sulphuric and nitric acids are the main mineral acids.

mineral-laden: a solution close to saturation.

mixture: a material that can be separated out into two or more substances using physical means.

molecule: a group of two or more atoms held together by chemical bonds.

monoclinic system: a grouping of crystals that look like double-ended chisel blades.

monomer: a building block of a larger chain molecule ("mono" means one, "mer" means part).

mordant: any chemical that allows dyes to stick to other substances.

native metal: a pure form of a metal, not combined as a compound. Native metal is more common in poorly reactive elements than in those that are very reactive.

neutralisation: the reaction of acids and bases to produce a salt and water. The reaction causes hydrogen from the acid and hydroxide from the base to be changed to water. For

example, hydrochloric acid reacts with sodium hydroxide to form common salt and water. The term is more generally used for any reaction where the pH changes towards 7.0, which is the pH of a neutral solution.

neutron: a particle inside the nucleus of an atom that is neutral and has no charge.

noncombustible: a substance that will not burn.

noble metal: silver, gold, platinum, and mercury. These are the least reactive metals.

nuclear energy: the heat energy produced as part of the changes that take place in the core, or nucleus, of an element's atoms.

nuclear reactions: reactions that occur in the core, or nucleus of an atom.

nutrients: soluble ions that are essential to life.

octane: one of the substances contained in fuel.

ore: a rock containing enough of a useful substance to make mining it worthwhile.

organic acid: an acid containing carbon and hydrogen.

organic substance: a substance that contains carbon.

osmosis: a process where molecules of a liquid solvent move through a membrane (filter) from a region of low concentration to a region of high concentration of solute.

oxidation: a reaction in which the oxidising agent removes electrons. (Note that oxidising agents do not have to contain oxygen.)

oxide: a compound that includes oxygen and one other element.

oxidise: the process of gaining oxygen. This can be part of a controlled chemical reaction, or it can be the result of exposing a substance to the air, where oxidation (a form of corrosion) will occur slowly, perhaps over months or years.

oxidising agent: a substance that removes electrons from another substance (and therefore is itself reduced).

ozone: a form of oxygen whose molecules contain three atoms of oxygen. Ozone is regarded as a beneficial gas when high in the atmosphere because it blocks ultraviolet rays. It is a harmful gas when breathed in, so low level ozone, which is produced as part of city smog, is regarded as a form of pollution. The ozone layer is the uppermost part of the stratosphere.

pan: the name given to a shallow pond of liquid. Pans are mainly used for separating solutions by evaporation.

patina: a surface coating that develops on metals and protects them from further corrosion.

percolate: to move slowly through the pores of a rock.

period: a row in the Periodic Table.

Periodic Table: a chart organising elements by atomic number and chemical properties into groups and periods.

pesticide: any chemical that is designed to control pests (unwanted organisms) that are harmful to plants or animals.

petroleum: a natural mixture of a range of gases, liquids and solids derived from the decomposed remains of plants and animals.

pH: a measure of the hydrogen ion concentration in a liquid. Neutral is pH 7.0; numbers greater than this are alkaline, smaller numbers are acidic.

phosphor: any material that glows when energized by ultraviolet or electron beams such as in fluorescent tubes and cathode ray tubes. Phosphors, such as phosphorus, emit light after the source of excitation is cut off. This is why they glow in the dark. By contrast, fluorescors, such as fluorite, emit light only while they are being excited by ultraviolet light or an electron beam.

photon: a parcel of light energy.

photosynthesis: the process by which plants use the energy of the Sun to make the compounds they need for life. In photosynthesis, six molecules of carbon dioxide from the air combine with six molecules of water, forming one molecule of glucose (sugar) and releasing six molecules of oxygen back into the atmosphere.

pigment: any solid material used to give a liquid a colour.

placer deposit: a kind of ore body made of a sediment that contains fragments of gold ore eroded from a mother lode and transported by rivers and/or ocean currents.

plastic (material): a carbon-based material consisting of long chains (polymers) of simple molecules. The word plastic is commonly restricted to synthetic polymers.

plastic (property): a material is plastic if it can be made to change shape easily. Plastic materials will remain in the new shape. (Compare with elastic, a property where a material goes back to its original shape.)

plating: adding a thin coat of one material to another to make it resistant to corrosion.

playa: a dried-up lake bed that is covered with salt deposits. From the Spanish word for beach.

poison gas: a form of gas that is used intentionally to produce widespread injury and death. (Many gases are poisonous, which is why many chemical reactions are performed in laboratory fume chambers, but they are a byproduct of a reaction and not intended to cause harm.)

polymer: a compound that is made of long chains by combining molecules (called monomers) as repeating units. ("Poly" means many, "mer" means part).

polymerisation: a chemical reaction in which large numbers of similar molecules arrange themselves into large molecules, usually long chains. This process usually happens when there is a suitable catalyst present. For example, ethene reacts to form polythene in the presence of certain catalysts.

porous: a material containing many small holes or cracks. Quite often the pores are connected, and liquids, such as water or oil, can move through them.

precious metal: silver, gold, platinum, iridium, and palladium. Each is prized for its rarity. This category is the equivalent of precious stones, or gemstones, for minerals.

precipitate: tiny solid particles formed as a result of a chemical reaction between two liquids or gases.

preservative: a substance that prevents the natural organic decay processes from occurring. Many substances can be used safely for this purpose, including sulphites and nitrogen gas.

product: a substance produced by a chemical reaction.

protein: molecules that help to build tissue and bone and therefore make new body cells. Proteins contain amino acids.

proton: a positively charged particle in the nucleus of an atom that balances out the charge of the surrounding electrons

pyrite: "mineral of fire". This name comes from the fact that pyrite (iron sulphide) will give off sparks if struck with a stone.

pyrometallurgy: refining a metal from its ore using heat. A blast furnace or smelter is the main equipment used.

radiation: the exchange of energy with the surroundings through the transmission of waves or particles of energy. Radiation is a form of energy transfer that can happen through space; no intervening medium is required (as would be the case for conduction and convection).

radioactive: a material that emits radiation or particles from the nucleus of its atoms.

radioactive decay: a change in a radioactive element due to loss of mass through radiation. For example uranium decays (changes) to lead.

radioisotope: a shortened version of the phrase radioactive isotope.

radiotracer: a radioactive isotope that is added to a stable, nonradioactive material in order to trace how it moves and its concentration.

reaction: the recombination of two substances using parts of each substance to produce new substances.

reactivity: the tendency of a substance to react with other substances. The term is most widely used in comparing the reactivity of metals. Metals are arranged in a reactivity series.

reagent: a starting material for a reaction.

recycling: the reuse of a material to save the time and energy required to extract new material from the Earth and to conserve non-renewable resources.

redox reaction: a reaction that involves reduction and oxidation.

reducing agent: a substance that gives electrons to another substance. Carbon monoxide is a reducing agent when passed over copper oxide, turning it to copper and producing carbon dioxide gas. Similarly, iron oxide is reduced to iron in a blast furnace. Sulphur dioxide is a reducing agent, used for bleaching bread.

reduction: the removal of oxygen from a substance. See also: oxidation.

refining: separating a mixture into the simpler substances of which it is made. In the case of a rock, it means the extraction of the metal that is mixed up in the rock. In the case of oil it means separating out the fractions of which it is made.

refractive index: the property of a transparent material that controls the angle at which total internal reflection will occur. The greater the refractive index, the more reflective the material will be.

resin: natural or synthetic polymers that can be moulded into solid objects or spun into thread.

rust: the corrosion of iron and steel.

saline: a solution in which most of the dissolved matter is sodium chloride (common salt).

salinisation: the concentration of salts, especially sodium chloride, in the upper layers of a soil due to poor methods of irrigation.

salts: compounds, often involving a metal, that are the reaction products of acids and bases. (Note "salt" is also the common word for sodium chloride, common salt or table salt.)

saponification: the term for a reaction between a fat and a base that produces a soap.

saturated: a state where a liquid can hold no more of a substance. If any more of the substance is added, it will not dissolve.

saturated solution: a solution that holds the maximum possible amount of dissolved material. The amount of material in solution varies with the temperature; cold solutions can hold less dissolved solid material than hot solutions. Gases are more soluble in cold liquids than hot liquids.

sediment: material that settles out at the bottom of a liquid when it is still.

semiconductor: a material of intermediate conductivity. Semiconductor devices often use silicon when they are made as part of diodes, transistors or integrated circuits.

semipermeable membrane: a thin (membrane) of material that acts as a fine sieve, allowing small molecules to pass, but holding large molecules back.

silicate: a compound containing silicon and oxygen (known as silica).

sintering: a process that happens at moderately high temperatures in some compounds. Grains begin to fuse together even through they do not melt. The most widespread example of sintering happens during the firing of clays to make ceramics.

slag: a mixture of substances that are waste products of a furnace. Most slags are composed mainly of silicates.

smelting: roasting a substance in order to extract the metal contained in it.

smog: a mixture of smoke and fog. The term is used to describe city fogs in which there is a large proportion of particulate matter (tiny pieces of carbon from exhausts) and also a high concentration of sulphur and nitrogen gases and probably ozone.

soldering: joining together two pieces of metal using solder, an alloy with a low melting point.

solid: a form of matter where a substance has a definite shape.

soluble: a substance that will readily dissolve in a solvent.

solute: the substance that dissolves in a solution (e.g. sodium chloride in salt water).

solution: a mixture of a liquid and at least one other substance (e.g. salt water). Mixtures can be separated out by physical means, for example by evaporation and cooling.

solvent: the main substance in a solution (e.g. water in salt water).

spontaneous combustion: the effect of a very reactive material beginning to oxidise very quickly and bursting into flame.

stable: able to exist without changing into another substance.

stratosphere: the part of the Earth's atmosphere that lies immediately above the region in which clouds form. It occurs between 12 and 50 km above the Earth's surface.

strong acid: an acid that has completely dissociated (ionised) in water. Mineral acids are strong acids.

sublimation: the change of a substance from solid to gas, or vica versa, without going through a liquid phase.

substance: a type of material, including mixtures.

sulphate: a compound that includes sulphur and oxygen, for example, calcium sulphate or gypsum.

sulphide: a sulphur compound that contains no oxygen.

sulphite: a sulphur compound that contains less oxygen than a sulphate.

surface tension: the force that operates on the surface of a liquid, which makes it act as though it were covered with an invisible elastic film.

suspension: tiny particles suspended in a liquid.

synthetic: does not occur naturally, but has to be manufactured.

tarnish: a coating that develops as a result of the reaction between a metal and substances in the air. The most common form of tarnishing is a very thin transparent oxide coating.

thermonuclear reactions: reactions that occur within atoms due to fusion, releasing an immensely concentrated amount of energy.

thermoplastic: a plastic that will soften, can repeatedly be moulded it into shape on heating and will set into the moulded shape as it cools.

thermoset: a plastic that will set into a moulded shape as it cools, but which cannot be made soft by reheating.

titration: a process of dripping one liquid into another in order to find out the amount needed to cause a neutral solution. An indicator is used to signal change.

toxic: poisonous enough to cause death.

translucent: almost transparent.

transmutation: the change of one element into another.

vapour: the gaseous form of a substance that is normally a liquid. For example, water vapour is the gaseous form of liquid water.

vein: a mineral deposit different from, and usually cutting across, the surrounding rocks. Most mineral and metal-bearing veins are deposits filling fractures. The veins were filled by hot, mineral-rich waters rising upwards from liquid volcanic magma. They are important sources of many metals, such as silver and gold, and also minerals such as gemstones. Veins are usually narrow, and were best suited to hand-mining. They are less exploited in the modern machine age.

viscous: slow moving, syrupy. A liquid that has a low viscosity is said to be mobile.

vitreous: glass-like.

volatile: readily forms a gas.

vulcanisation: forming cross-links between polymer chains to increase the strength of the whole polymer. Rubbers are vulcanised using sulphur when making tyres and other strong materials.

weak acid: an acid that has only partly dissociated (ionised) in water. Most organic acids are weak acids.

weather: a term used by Earth scientists and derived from "weathering", meaning to react with water and gases of the environment.

weathering: the slow natural processes that break down rocks and reduce them to small fragments either by mechanical or chemical means.

welding: fusing two pieces of metal together using heat.

X-rays: a form of very short wave radiation.

Index

acid 12, 21, 29, 39, 42
acid rain 13
adhesive 22, 23
algae 4, 40
ammonite 9
anion 29
anode 37
antacid 5, 42, 49
atmosphere 12
atomic number 44, 47

base 21, 25
body 5, 32
bond 49
bone 5, 32, 33
building stone 18
burnt lime 22

Ca 4, 44
calcite 8, 10, 14, 16, 33
calcium 4, 6, 38, 40, 42, 44
calcium bicarbonate 15, 27, 38, 42
calcium carbonate 4, 8, 12, 15, 16, 21, 26, 33, 39, 43
calcium fluoride 8
calcium hydrogen carbonate 38
calcium hydroxide 6, 24, 26, 28
calcium ions 25, 28, 40
calcium oxide 20, 24, 48
calcium phosphate 32
calcium sulphate 8, 30
carbon dioxide 12, 19, 26, 43, 49
carbonic acid 12
cathode 37
cation 29
caustic 22
caves, cave formations and caverns 14
cement 5, 21, 22
chalk 4, 10
chlorophyll 34
clay 29
colour 8, 10
compound 4, 7, 41
concrete 5, 23
copper sulphate 37
coral 4, 9, 10
corrosion 13, 36
crystals 8

desert roses 8
detergent 40
dissolve 12
dolomite 5, 37

electrode 37
electrons 47
element 4
equations 48–49

flocculate 25
flowstone 14
fluorspar 8
fossil 11
fossil fuels 13

geysers 17
glass 22
gypsum 8, 30

hard water 38, 40
hot springs 17
hydrated, lime 24
hydrochloric acid 34, 43
hydrogen 7, 34
hydroxide ions 25, 29

Iceland Spar 8
ion 29, 41
iron 36

karst 11

lather 38
lava 8
lime 22, 24
limescale 16, 38
limestone 4, 10, 12, 19, 20
limewater 7, 26

magnesium 5, 34–37, 38, 45
magnesium carbonate 37, 43
magnesium hydroxide 42, 43
magnesium ions 34, 40
magnesium oxide 37
magnesium ribbon 34, 37
marble 4, 18
Mendeleev, Dmitry Ivanovich 46
metal 6, 44, 45
Mg 5, 45
Milk of Magnesia 42
mineral 8, 15
mortar 22

neutralise 21, 25, 43

oil storage tanks 37
oolitic limestone 10
ore 10
organic acids 39
oxidation 35

pearl 4
percolate 14
Periodic Table 46–47
permanent hardness 38
phosphorus 40
plaster 5, 31
Plaster of Paris 31
pollution 13
porous 10
Portland cement 23
precipitate 7, 16, 26, 38
precipitation 14
protons 47

quicklime 20, 22, 24

reaction 6
reactivity series 36
Reims cathedral 13
resin 41
rhombohedral-shaped crystal 9

saturated 6
scale 10, 39
scum 40
sink holes 12
sinter 17
slaked lime 24, 28
sodium ions 40
soil 12, 24, 28
soil conditioner 24, 28
solution 12, 16, 27
stalactite 14
stalagmite 14, 39, 44
steam 24

swallow holes 12

teeth 5, 33
temporary hardness 38
tin 36
travertine 8, 14, 16

wallboard 30
water 38, 40
water-softener 40
weathering 12, 19
whitewash 23